T0146128

Blip, Ping & Buzz

MARK DENNY

..........Making Sense of Radar and Sonar

THE JOHNS HOPKINS UNIVERSITY PRESS
Baltimore

The Johns Hopkins University Press
2715 North Charles Street
Baltimore, Maryland 21218-4363
www.press.jhu.edu

Library of Congress Cataloging-in-Publication Data
Denny, Mark, 1953–
    Blip, ping and buzz: making sense of radar and sonar
      p. cm.
    Includes bibliographical references and index.
    ISBN 13: 978-0-8018-8665-2 (hardcover : alk. paper)
    ISBN 10: 0-8018-8665-1 (hardcover : alk. paper)
    1. Radar—Popular works.   2. Sonar—Popular works.
    3. Remote sensing—Popular works.   4. Signal processing—
    Popular works.   I. Title.
    TK6576.D46 2007
    621.3848—dc22   2006101449

A catalog record for this book is available from the British Library.

To Bill Denny and his brother:

two engineers who tried to teach me

the value of practical things

Contents

Acknowledgments

I am grateful to the following individuals and organizations for providing me with excellent images: Prof. Brock Fenton, Captain Jerry Mason, Barry Stewart, Stephen Wilkinson; The Battle of Britain Historical Society (United Kingdom), Institute for Ocean Sciences (Canada), Klein Associates (United States), NASA, Shannon Dolphin and Wildlife Foundation (Ireland), U.S. Air Force, U.S. National Archives, U.S. Navy. For helping to create this book, I thank Trevor Lipscombe, Martha Sewall, and Nancy Wachter of the Johns Hopkins University Press.

Blip,
Ping
& Buzz

Introduction..............

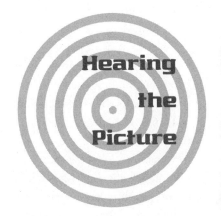

Hearing the Picture

Every author should explain to his readers, up front, what the book they hold in their hands is about and what it is not about. After all, if you are to invest some hard-earned cash in buying this book and a lot of time in reading it, then you naturally want to know whether the subject is of interest. The quick explanation, expanded over the next few paragraphs, is this: my book describes the ideas behind radar and sonar signal processing. I will use nonmathematical language (as far as possible) to explain the techniques of image processing and the physics behind human and nonhuman remote sensing.

Certain whales and dolphins can *echolocate*, as can a large family of bats; this means they provide nature's contribution to remote sensing. Humans have developed the rest: *radar* systems based on land and on surface ships, and inside civilian and military aircraft; *sonar* systems housed in ships and submarines; and both radar and sonar in remotely operated vehicles. "Blip," in the title, is the engineer's word for the pulse sent out by a radar transmitter. "Ping" is the sonar engineer's equivalent, and "buzz" is the noise made by echolocating bats.

The most advanced remote-sensing system today is what I'll call the MB800 family of airborne sonar processors (described in chapter 5). These machines have a processing capability that in some ways exceeds that of better-known sensors, such as the massive Airborne Warning and Control

System (AWACS) "eye in the sky" airborne radar system, yet the MB800 central processor weighs less than half a gram, and a whole airframe can fit in a cigarette pack. The capabilities of the MB800s, and of the much cruder early radars (such as the Chain Home system that saved Britain during World War II, described in chapter 1) are outlined here and placed in historical and technological context.

Remote sensing is a huge subject. It would be foolish of me to try to cover every aspect of it in a single book. My purpose is to get across the core ideas in a meaningful way, rather than equip you with a degree in electrical engineering, aeronautics, or bioacoustics. So, I must carefully cherry-pick the topics to be elucidated, and ruthlessly suppress the rest. Given that I am not a biologist, I won't attempt to describe the detailed biological adaptations of the organic members of our remote-sensing fraternity. Bats and whales are treated as sonar signal processors. They are part of a list of remote-sensing machines: submarine sonar, whale echolocation, land- and ship-based radar, airborne interceptor radar, and bat interceptor sonar. My only concession is terminological since, for example, it seems silly to apply engineering terms like "directional antennas" to a bat, when "ears" describe the same thing more clearly and succinctly. In the same vein, I will brutally suppress any description of remote-sensing electronics, for two reasons. First, my purpose is to enlighten you about the principles and key ideas that underpin remote sensing, not to lecture you about how engineers implement the ideas in hardware.[1] Second, despite an electronic engineering career spanning two decades, I know little about electronics. I worked as a systems engineer for multinational aerospace corporations designing radar and sonar signal processors, so my experience and expertise reside in the ideas and techniques, rather than in the practical details of their realization.[2]

---

[1]If you like, this book is about the software of remote sensing, not the hardware implementation. It is the *algorithms* of radar and sonar signal processing that interest me; I do not care much whether these algorithms are embedded in silicon or in DNA, whether the processor is housed in a carbon-composite airframe or in a fur-covered skull, or whether the interceptor target is a missile or a moth.

[2]Real engineers, whether electronic ("sparkies," in the Scottish vernacular of my erstwhile colleagues) or mechanical ("clankies"), often despair at the fanciful and elaborate schemes that their systems colleagues design to solve a given problem. An elegant mathematical solution is useless if it cannot be practically implemented, so we systems engineers always have to keep at least one foot on the ground. I will have something to say about the social dynamics of large organizations that seek to build complex signal processors. For now I note simply that a number of technical disciplines must work together to make it all happen, and sometimes (always?) there is tension across boundaries. I recall one electronics engineer colleague, sounding like Scotty from *Star Trek*, exploding at a mechanical engineer: "If it's no covered in oil, you clankies dinnae understand it!"

I do not intend to write a textbook, though this book may well be read as a supplementary text. I am guessing that you are a technology buff who has wondered from time to time how radar or sonar works. Or perhaps you are a student who needs to know what remote sensors can do, but not every last detail of how they do it. You do not have the time or inclination to work through a degree in radar engineering, but you have been left unsatisfied by the comic-book explanations occasionally proffered in newspapers. You probably don't want to read several hundred pages of dense mathematics (and, believe me, a detailed treatment of signal processing gets very mathematical, very quickly). Anyway, the subject matter of remote sensing is cross-disciplinary, so I would need to write several academic tomes to cover it all, on optics and acoustics, engineering and signal processing, mathematics and statistics. So from the beginning I, as author, am faced with a conundrum: how can I convey to you a very technical subject in nonmathematical language, without jettisoning the central ideas? Also, some of you will harbor unreasonable prejudices against, say, statistics, considering it to be dry as dust or dull as ditchwater.

Happily, there is a way for me to get across to you the important ideas that underpin remote sensing without being dull or excruciatingly technical. The result is, though I say so myself, a unique read. I am unaware of any other book that broaches our subject with a light yet accurate touch. I feel like the chef who is pleased with himself (puffed up like one of his pastries, you might say), having created an airy yet substantial soufflé, tasty and satisfying, fluffy but not frivolous. Enough bragging, I hear you mutter: tell us how. Well, remote sensing is unique in that many of the fundamental ideas can be conveyed with a few simple diagrams. The math behind such diagrams may be (often is) horrendous, but the core idea can be conveyed with clarity without math. I know of no other technical subject where so much of the foundations can be elucidated by this approach. My guess as to why remote sensing lends itself to a pictorial explanation? It is at bottom a geometrical subject. Even the signal processing is geometrical, since the aim of remote sensing is to form an image, a picture. So, in the following pages you will find many diagrams; these combined with a few paragraphs of text will get across complex ideas, such as superposition and aliasing, correlation and integration, phased arrays and beamforming. From these bricks, we can build the cathedrals of AWACS and MB800, as you will see.

There are a few mathematical equations in this book, but not many. Those of you who shy away from equations as horses shy from rattlesnakes need not despair: you will be warned in advance when an equation is lurking in the scrub, and these equations will have their stings drawn—they will be tame and friendly. Cuddly, even. To be entertaining as well as informative, I have also chosen with care a number of the examples to illustrate a given point. There is *no excuse* for an author who writes a dull page about, say, radar clutter, when so many amusing, historically relevant, and informative examples are at hand. Read on.

Seeing and Hearing

Remote sensing is a way of looking at the world—literally—using electromagnetic radiation (microwaves, in the case of radar) or sound waves (sonar). The word *radar* is an acronym for *r*adio *d*etection *a*nd *r*anging, and *sonar* is an acronym for *so*und *na*vigation and *r*anging.[3] The radiation is transmitted through air or water, bounces off a *target*,[4] and is reflected back to a receiver. The wonders of signal processing then turn the jumble of received signals into an image. That is the essence of remote sensing.

The transmitted radiation consists of electromagnetic (EM) waves for radar and sound waves for sonar. These waves are very different in nature, but the signal processing of them is similar; many of the techniques used in forming radar images are also applied in sonar signal processors. The crucial factor, as we will see, is that the radiation consists of waves. The EM spectrum spreads from very short wavelength cosmic rays to very long wavelength, low-frequency radio waves, and includes visible light, infrared and ultraviolet waves, and microwaves. For reasons I will explore later on, only a small fraction of this spectrum is suitable for remote sensing. Most modern radars use microwaves, with wavelengths typically 0.3–100 cm (1 inch = 2.54 cm). An exception is the laser range finder, which utilizes infrared wavelengths. The wavelengths used by sonar transmitters—in air and water—are usually much shorter (i.e., higher frequency) than those we humans can hear.

---

[3]These acronyms are now pretty much universal, though it was not always so. Reflecting the diverse origins of human remote sensing, the early names varied from country to country. Thus, the English originally called radar *RDF* (radio direction finder) and designated sonar *ASDIC* (after the wartime Allied Submarine Detection Investigation Committee that oversaw its development). In biology the descriptive label *echolocation* was introduced to describe sonar in bats and whales, soon after humans realized that we were not the first to develop remote sensing.

[4]This is the first of many technical terms that you will need. (For convenience, all are included in the Glossary; when first introduced in the text they are *italicized*.) A target need not be something you want to destroy (for example, a civilian aircraft), but on the other hand it might be (for example, an incoming missile). A target is the object that you wish to locate with your remote sensor.

Given that radar makes use of EM radiation, whereas sonar utilizes sound, it would seem natural to think of radar as "seeing" and sonar as "hearing," but this is a rather superficial analogy. By and large, I will use in this book a different analogy connecting remote sensors with human senses. First, two more technical terms. A *passive* sensor is one that makes use of natural radiation (light or sound) that arises from the surrounding environment, whereas an *active* sensor supplies its own radiation, which is transmitted and then detected. With this distinction, which is a basic one in remote sensing, we can classify human sight as passive sensing, since it utilizes the visible light emitted by the sun. We do not emit our own light to see by. Hearing is different. True, we make use of sounds emitted by others and by the natural environment, and this is indeed passive sensing, but we also transmit acoustic radiation (in plain words: we speak) and listen for a reply. In this sense hearing is active. So I draw an analogy between passive sensors and seeing, and between active sensors and hearing.[5] If you buy this analogy, then radars are hearing aids. After all, we talk about a radar *echo*, and about electronic *noise*, which suggests listening. All active sonar and echolocation is listening. On the other hand, a weather satellite that monitors EM radiation from the earth is *looking* at the earth, and a secret service station that monitors the electronic transmissions across the world is *seeing* them.

The distinction between active and passive sensing is basic because of the enormous advantages that accrue only to active sensing, which is why our radar and sonar sensors are mostly active. An active sensor transmits a known signal, of known power and frequency, and of known *waveform* (of which much more later), and so it knows what to listen for. This advantage means that an active sensor can be spectacularly sensitive; as you will see, it can detect echo signals that are billions and billions of times fainter than those it transmitted, and so it can hear a very long way. A radar can hear where the moon is, to within a fraction of an inch. A bat can hear where a bug is, and can form a very high-resolution picture from what its ears pick up. I might have called this book "Hearing the Picture." It means processing the signals transmitted and received by an active sensor.

From Chain Home to MB888

Later I will share with you what I know about the mighty MB800 family of airborne acoustic processors, but I'll begin the book with a much humbler

---

[5]So, according to my analogy, you *see* birdsong with your ears, since this is passive sensing. When you use a flashlight to find your way around in the dark, you are *hearing* with your eyes.

yet historically more important remote sensor system that dates back to the early days of radar. In between you will learn about how the processing is done, about how signals propagate through the air or water, and about how the targets try to be silent—undetected—if they fear that your radar/sonar wants them destroyed. (Some want to be heard: I will show you how civilian aircraft make themselves known to air traffic control, or ATC, radar.) The development of human remote sensing is less than a century old, and I'll place the early days of this development in historical context. The story is an interesting and historically important one, which explains why radar and sonar signal processing is a multibillion-dollar industry today, across the world, employing tens of thousands of highly trained engineers, mathematicians, technicians, ergonomists, physicists, geologists, industrialists, computer scientists, archaeologists, biologists, and a few paleontologists. Remote sensing has found applications in the military, primarily, seeking out and tracking the enemy in the air, on the ground, and underwater. It has both military and civilian applications onboard satellites in space, both looking (actually, listening) outward to the solar system and inward to the earth below. Remote sensors listen to the weather, or penetrate through it to see (hear) the enemy aircraft hoping to hide behind clouds. Remote sensors seek out shoals of fish, as we will see, or dismiss the clatter made by a trillion copulating shrimp[6] to detect the submarine lurking behind. Remote sensors track missiles and moths, battle tanks and bugs, Navy SEALs and, well, seals. And they all use basically the same processing techniques. These techniques, at least those that we know of, are the main technical subject tackled in this book.

Did I really say paleontologists? Indeed. Human remote sensing may be only a few decades old, but bats and whales have been doing it for tens of millions of years, as fossil remains show. Bat and whale echolocation is also a significant part of the story I wish to tell. Which brings me to the MB800 family of ultra-high-tech, supercomputer-powered, most advanced remote sensors on the face of the earth today. Some of you may already have guessed that I am referring to bat echolocation and, in particular, to the 800 or so species of microchiropteran bats. For those who did not, I apologize for the slight deception—giving these little creatures a pseudo-military designation and describing them in terms of machines—but I did so to make a point. Bat signal processing is more advanced than ours, despite our multibillion-

---

[6]Really. To a sonar receiver this noise source is a considerable, albeit seasonal, nuisance. Take my word for it.

dollar research programs. They can do things we cannot; we have no idea how bats form their sound pictures,[7] but we know that they do form such pictures, and that they form them in a manner quite different from anything that human ingenuity has yet devised. An eloquent admission of our subordinate role in remote sensing is provided by the fact that the U.S. military spends part of its remote-sensing research budget on biological studies that seek to understand how bats do it. Humans are catching up, however, and in the last chapter I outline some of the more advanced signal processing techniques being developed today.

One final point. Because they cast interesting sidelights on the story I wish to tell, please be sure to read the footnotes.

---

[7]This is not quite true. In chapter 5 I will lay before you the experimental evidence and the considerable amount of "reverse engineering" inferences made by biologists and engineers to understand what makes bat echolocation so good.

1 ....  Early Days

Let's begin with a look at the frantic early days of radar, immediately before the Second World War and during the first couple of years of that epoch-changing conflict. In particular, we'll focus on the story of the first radar application: the clunky but ultimately successful Chain Home (*CH*) early warning system that proved to be the decisive factor during the Battle of Britain in 1940. My aim is not to provide a comprehensive technical history of radar—there are already many good and excellent books in this area, to which I will refer—but rather to use the Battle of Britain as sugarcoating for a "Radar 101" pill. You will read some interesting history while learning a lot about the basic concepts that we will need later on. The technical sections (usually a picture with a few paragraphs of explanation) will be separated from the main story, for continuity, and appear at the end of the book. Pointers to each topic will pop up within the relevant history section. Thus, when I talk about the earliest wave-interference radar devices, there will be a pointer to note T1, a short and painless aside describing wave interference. In the same vein, I will not discuss beamforming until the subject arises in the main text. I say "main text" to describe the historical component of this chapter, but you may wonder at first what is "main" about it, since there will seem to be more of the technical stuff. I guess this is inevitable, because I have to do a lot of spadework before the foundations go in, next chapter. As I said in the introduction, this is a huge subject.

The Invention of Radar

In World War II more resources were allocated to radar development than to any other technical project, except for the development of the atom bomb. This stark fact neatly summarizes the military importance attached to radar during its early years. One man who played a significant role in this development was Lee DuBridge, a prominent American physicist and academic administrator, who said "The bomb may have ended the war, but radar won the war." The military significance of radar continues to this day. Given the importance of radar to military success it is hardly surprising that, retrospectively, the history of radar has been tinged with more than a little national bias. A German technical history of radar development credits Germans with many of the key developments, whereas U.S. historians concentrate on the (considerable) American effort. Most Englishmen think that radar is an English invention, whereas most Scots will tell you that the British push to develop the first radar system was under the guidance of a Scotsman, not an Englishman. A French history emphasizes the prior claims of—you guessed it—a Frenchman. In fact Germans, Americans, British, French, Swiss, Russians, Italians, Japanese, and others all toyed with the idea of radar in the 1930s, quite independently of one other. Some of the national efforts were half-hearted or leisurely, while others were frantic and desperate, depending on the circumstances, as we will see. The potential military importance of "radio detection" was appreciated by all, however, which explains why each nation conducted its research and development programs secretly, or at least away from the searching gaze of other nations. This rule of independent development would be suddenly and quite astonishingly broken by the Tizard technical mission of 1940. But I am getting ahead of the story.

Perhaps all the claims for the invention of radar are plausible, given the independent efforts of different groups.[1] The more objective historians of radar recognize this fact. French academician Maurice Ponte has asserted the uncertain parentage of radar as follows: "The fundamental principle of the radar belongs to the common patrimony of the physicists," (Wikipedia) whereas the American radar guru Merrill Skolnik, author of a standard textbook read by all students of radar engineering, expresses the same sentiment

---

[1]Sometimes, even within national borders, different groups would pursue their research independently. Thus, in the United States, the Army Signals Corps and the Naval Research Lab did their own thing. In Germany also, rivalry between the different armed forces hindered collaborative R&D efforts.

more economically, referring to the "many fathers of radar." In this section I will concentrate on the key developments—deployment rather than invention—that occurred prior to the Second World War.

The Scottish physicist James Clerk Maxwell first explained electromagnetic (EM) radiation in the 1860s. He showed that it could travel through a vacuum, which is not possible for the other known types of waves (such as sound waves, water waves, or shock waves) and that the speed of all EM waves is the same as the speed of light.[2] EM radiation includes radio waves, which have very long wavelengths and low frequencies, and gamma rays with very short wavelengths and high frequencies. In between are found (in order of increasing frequency) microwaves, infrared, visible light, ultraviolet, and X-rays. Our eyes are sensitive to only a narrow band of EM radiation, and we call that portion of the EM spectrum *light*. A generation after Maxwell, the German physicist Heinrich Hertz performed experiments with radio waves, and showed that they were transmitted through certain materials and were absorbed by or reflected off other materials. Hertz[3] was the first person to generate and detect radio waves (in 1887).

The next step toward radar was taken by another German, in 1903. Christian Hülsmeyer of Dusseldorf experimented with radio waves reflected off ships at sea. He showed that it was possible for one ship to detect the presence of another ship using radio waves, at night or in fog when the two ships could not see each other. He conducted public demonstrations in Germany and in Holland, proposing his system as a collision avoidance device for use in harbors. When a ship was detected (this equipment had a maximum range of 3 km—say 2 miles), a bell was rung. Hülsmeyer patented his device in several countries in 1904. Unfortunately, nobody was interested in producing his system or in sponsoring further development, and so it withered on the vine. As a historical footnote, though, we can say that, thanks to Herr Hülsmeyer, radar has a slightly more ancient

---

[2]The speed of light is generally denoted by the letter c, and in a vacuum it has a value of three hundred million meters per second ($c = 3 \times 10^8$ m s^{-1} in scientific notation) or 186,000 miles per second. Nothing can go faster than this speed and, according to Einstein, c is independent of the speed of the source. In other words, if a star is heading toward you at speed $v = 1,000$ m s^{-1}, then you would measure the speed of light emitted by that star as c, not $c + v$ as you and Isaac Newton may think. Don't blame me for this weird fact—blame Einstein. In a medium such as air or glass, the speed of light is reduced from c to c/n, where n (the *refractive index* of the medium) exceeds 1. For glass typically $n = 1.5$, whereas for air $n = 1.000277$.

[3]A unit of frequency is named after him. One Hertz (1 Hz) corresponds to one cycle per second. One kilohertz (1 kHz) is 1,000 cycles per second. One Megahertz (1 MHz) is 1,000,000 cycles per second.

lineage than (human) sonar, which dates from 1912 following the *Titanic* disaster.

Nikola Tesla, the great Serbian-American inventor who made many contributions to electrical engineering, clearly understood the principles of radar in 1917: "we may produce at will, from a sending station, an electrical effect in any particular region of the globe [to] determine the relative position and course of a moving object, such as a vessel at sea, the distance traversed by the same, or its speed." Later, in the early 1930s, Émile Girardeau in France developed radar, "conceived according to the principles stated by Tesla." Guglielmo Marconi, the Italian-British engineer who was the first to transmit a wireless signal across the Atlantic (and later won a Nobel Prize in physics), also foresaw radar, and in a 1922 speech stated the principles of remote object detection by short-wavelength radio waves. By 1933 he had developed a working device. Other innovators in other nations also were on the trail and so, in the 1920s, the race for a working radar system began. The race picked up pace in the 1930s as the political situation in Europe deteriorated. Aircraft were becoming an important military factor, and many people in military and industrial circles recognized the need to defend against them.[4] At the same time radio technology was improving, and so along with the dragon came the possibility of a sword to slay it.

Pre-war Development

In 1922 A. H. Taylor and L. C. Young of the Naval Research Lab (NRL) in the United States detected a ship by using CW interference radar. Here CW stands for continuous wave, meaning that the EM radiation is continuously transmitted and is picked up at a distant location by a receiver. The receiver has to be widely separated from a CW transmitter (a *bistatic* system, in modern radar parlance) because otherwise it would be overloaded by the transmitted signal. Imagine standing next to a foghorn, hoping to

---

[4]In Britain the first attempt at an early warning system to detect hostile aircraft was not radar, but passive sonar. In 1933 the War Office funded W. S. Tucker to construct large concrete acoustic mirrors at Romney marsh on the south coast of England. The largest such mirror was 200 feet long and 25 feet high, and pointed out to sea. Sensitive microphones were placed at the focal points of the mirror, to pick up the sound of incoming enemy planes. Smaller mirrors (a mere 30 feet long) were steerable, to provide an indication of aircraft direction. Unfortunately the detection range was only 15 miles, and then only under the most favorable conditions. Passing vehicles disrupted detection. The short detection range provided only 3 or 4 minutes warning. The acoustic mirror project was cancelled in 1936.

listen for an echo from the foghorn, reflected off a distant ship at sea. You would not be able to hear the faint echo if the nearby foghorn blasted away continuously—you would be deafened by it. So, CW systems have to be bistatic. Taylor and Young detected the ship by exploiting the phenomenon of *wave interference* (this and other fundamental wave characteristics and phenomena are explained in note T1). A signal (in this case the transmitter produced radio waves with a wavelength of 5 meters) travels from the transmitter to the ship, is reflected off the ship, and into the receiver. The transmitted signal also travels directly to the receiver. So, the receiver hears two signals: direct and reflected. The interference between these produces an amplitude-modulated signal.[5] With no reflection there is no interference and so no signal modulation. Thus, if the receiver outputs a modulated signal then you know that a ship (or some other reflector) must be present. The Taylor and Young apparatus tells *that* a ship is present, it does not tell *where* it is. Thus, their system was a detector, not a locator. All CW devices have this shortcoming; they may act as a trip wire, but nothing more. This defect was well recognized at the time. Taylor and Young submitted a proposal for further work, but their proposal was turned down.

The first British radiolocation patent was obtained in 1928. At this stage the British were well behind other nations, in particular, the United States. That situation would change in the next decade, as we will see, due to the increased political tensions in Europe and the perceived vulnerability to invasion. Consequently, greater effort and more urgency were put into British radar R&D during the 1930s than elsewhere.

A CW interference radar device first detected an aircraft in 1930. This was the result of a happy accident. In 1930 L. A. Hyland of the NRL was experimenting with a 33-MHz CW interference radar.[6] Two miles separated his transmitter and receiver, and between them lay an airfield. It just so happened that an airplane was landing at the same time as Hyland was operating his CW device, and this airplane caused discernible interference. Encouraged by this, and by the earlier success detecting ships, in 1933 Taylor, Young, and Hyland obtained a joint patent for their "system for de-

---

[5]This modulation due to a reflector is exactly what happens when an aircraft overhead causes your TV picture to flutter. Instead of receiving a single signal from the transmitter mast, your TV also picks up the transmitter signal reflected off the aircraft.

[6]Wave speed c is related to frequency f and wavelength λ via $c = f \lambda$. We have already seen that c is fixed at 300,000,000 m s^{-1}, and so a 33-MHz frequency yields a wavelength of about 9 m. Henceforth, I will simply state the wavelength, or the frequency, of an EM wave, since one can easily be calculated from the other, given that c is constant.

tecting objects by radio." The next year the NRL was able to demonstrate a practical 60-MHz (i.e., with a wavelength of 5 m) wave-interference radar.

Hitler rose to power, increasingly absolute, in Germany in 1933. In the same year the *Kriegsmarine* (German navy) begin research into *Funkmesstechnik* (remote radio measuring technology). Russian research began the following year. The pace of radar development was picking up across the world, and with it an expanding realization that pulsed transmissions (rather than CW) at microwave wavelengths (rather than the longer radio waves) were the way forward. Pulsed transmission radars permit an estimate of the aircraft or ship range,[7] note T2, and so convey more useful information than the trip-wire CW devices. Before and during World War II the radar systems deployed by both sides would be increasingly at shorter and shorter wavelengths, because these wavelengths provided better target position information and could detect smaller targets, for reasons that I will get to soon. Independent radar development also arose in Italy, France, Japan, and other countries: all were initially CW interference devices and all tended toward short-wavelength pulsed radar systems.

The year 1935 was an important one in the history of radar. In Germany the radio communications and microwave radar work of Hans Eric Hollmann was getting under way. A 50-cm wavelength ship-detection radar was shown to be able to detect marine vessels to a range of 10 km. A ship 8 km away was located with an accuracy of 50 m. This was considered to be good enough for gun laying. Aircraft at an altitude of 500 m were detected at a 28-km range. This development work would soon lead to two of the standard German radar systems used throughout the war, *Freya* and *Seetakt*. The year 1935 saw the establishment in England of Henry Tizard's Aeronautical Research Committee, the body that oversaw British radar development during the years 1935–1940. In 1935 Tizard asked Robert Watson-Watt (a Scot, and a descendent of James Watt, who developed the first high-pressure steam engine in 1780) to look into the possibility of developing an EM "death ray."[8] Watson-Watt concluded that such a ray would require so much power as to be totally impractical, but offered the sugges-

---

[7]Henceforth, I will use the technical term *target* to refer to the object sought by a radar system.

[8]The notion of invisible death rays was popular at the time, and gave rise to many crackpot inventions that were aired publicly. As an incentive, a prize of £1,000 was offered (by whom, I have not been able to unearth) for the first such ray that could be shown to kill a sheep at 100 yards. You will not be surprised to learn that no animals were hurt during the testing of these products, and the prize money was not collected.

tion that radio waves may be used for detection, rather than destruction. In a memo to the Air Ministry he proposed a method for "Detection and Location of Aircraft by Radio Methods." In what has become known in radar circles as the "Daventry experiment," named after a nearby town, Watson-Watt successfully demonstrated a receiver that detected aircraft out to a range of 13 km (8 miles) by using the wave-interference idea—in a manner similar to the earlier U.S. experiment; the transmitter was a 49-m wave-length civilian BBC radio station.[9] This led to full-scale development of a radar early warning system for the English coast. The same year, after intensive effort, his team demonstrated a pulsed radar that located aircraft out to 65 km (40 miles); this system was able to estimate the aircraft target range, and also altitude (from the elevation angle; see note T3). Finally, in 1935 the French deployed a ship-collision-avoidance radar.

The first successful U.S. demonstration of a pulsed radar occurred in 1936, in experiments conducted by Robert M. Page. The initial detection range for an aircraft target was only 2.5 miles, but this was soon extended to 25 miles (40 km). The operating frequency of this system was 28.3 MHz, giving a wavelength of 10.6 m, and the pulse duration of $\tau = 5$ microseconds (5 μs) provided a range estimation accuracy of $\pm$ 750 m. This same year the British extended the range of their aircraft detection system to 150 km, and reduced the transmitted wavelength from 25 m to 12 m. In America the NRL developed a high-frequency (HF) 200-MHz (1.5-m wavelength) radar—note the trend toward shorter wavelengths.

In 1937 R. C. Newhouse of Bell Labs conducted experiments that led to the first radio altimeter. These instruments operate on the radar principle, and represent the first commercial (as opposed to strictly military) application of radar. This year also saw full military deployment of the German *Freya* and *Seetakt* radar systems, though in low numbers. *Freya* was a 2.4-m wavelength aircraft detection radar with a range of 80 km. *Seetakt* with 80-cm wavelength transmissions detected ships out to a 14-km range,[10] with an accuracy of a couple of degrees. By 1937 the British had constructed the first seven of their Chain Home radars along a section of the coast of south-

---

[9]Watson-Watt was so encouraged by the success of his experiment that, foreseeing the means to stop enemy bombers, he declared "Britain has become an island again!"

[10]You may well ask why aircraft could be detected out to 80 km but much larger ships could be seen only out to 14 km. The reason is that the radar transmitters in the late 1930s were very, very low powered at the shorter wavelengths. So why bother at all with short wavelengths? It turns out that (and I will show you in T4 how this works) shorter wavelengths give more accurate angle estimation, for a given antenna size—note the impressive *Seetakt* azimuth accuracy.

ern England. In September 1938 these radars tracked the aircraft that carried Prime Minister Neville Chamberlain to his ill-fated meeting in Munich with Hitler and Mussolini. That same month the CH system became operational 24/7/365 until the end of the war. Also in 1938 the U.S. Army Signals Corps first operated their SCR-268 fire-control radar; because of its impressive range accuracy, this system became standard equipment until 1944.

War in Europe began one year after Chamberlain's trip to Munich, in September 1939, by which time there were 19 CH radar systems linked together, strung along the south and east coasts of England, and organized into an efficient command and control structure. Just in time. In the United States during this period six SCR-270 long-range early warning radar sets were deployed at Pearl Harbor.

Chain Home

I will describe the ponderous and homely CH system, because it was the first radar system used in anger, and because it serves to introduce a lot of radar and signal processing topics. Also, the contribution CH made to the Battle of Britain is historically significant and is an interesting read.

CH radars are remembered fondly in British engineering circles in just the way that vintage cars bring tears to the eyes of their proud owners, who overlook the deficiencies of their pride and joy. Foreigners, more objectively, shake their heads and wonder how such a dog's breakfast was ever cobbled together to form a working air defense system. CH has been described as obsolete from the day it was first switched on. It made use of HF radiation (radio waves), not microwaves, with a wavelength of $\lambda = 12$ m. This choice of operating frequency/wavelength is a bad start for two reasons. First, for a given antenna size, long wavelengths provide wide beams and so achieve worse angular accuracy than short wavelengths. (This is true, in general, and is not just a quirk of CH radars. Note T4 gives you all you need to know about beamforming.) So the bearing of an incoming enemy aircraft cannot be determined as accurately with long wavelengths. Second, long wavelengths mean poor target resolution—the radar cannot separate planes or other targets that are closer together. (Angular resolution is not the same as angular accuracy: see note T5.) In fact, the CH system did tend to underestimate enemy numbers; the long operating wavelength may have been a contributory factor. So why did Watson-Watt and his team[11] choose

---

[11]His senior assistants on this and later wartime radar systems were A. F. Wilkins and E. G. Bowen.

long radio waves? It is because they needed large transmitter power to attain large detection ranges, and the technology of 1937–1939 included only feeble transmitters at short (microwave) wavelengths. Large detection ranges were a must; CH was able to detect the German bomber squadrons assembling over French airspace, prior to their heading northward across the English Channel with fighter escorts. This gave the Royal Air Force (RAF) enough time to scramble its Spitfire and Hurricane fighters and get to altitude before encountering the Germans. An altitude advantage was and still is crucial in any air combat—and getting caught on the ground is fatal. So, tactical necessity plus technological shortcomings dictated CH operational wavelength.

The CH radar has also been described as a "technical anomaly" and a "dead-end design." Both statements are true. A British technical account considers CH to be a stopgap measure; CH "held the fort" until high-power microwave transmitters became available in 1940.[12] All these sideswipes at CH, made with the benefit of hindsight, prepare us for a low-end, hastily assembled, gimcrack radar system. Indeed, the Chain Home system was hastily assembled. The always-pragmatic Watson-Watt said that "you never get the best device, and the second best comes too late," so he built the third best. This attitude was in marked contrast to the German approach that "the best is just good enough." Consequently, German radars were superior to British ones at the beginning of the war. And yet history tells us that the CH radar system emerged victorious. As American historian of radar technology Louis Brown said: "It is idle and mean-spirited to criticize something so obviously successful as the CH radar." So how on earth could the inferior radar succeed? The answer is an important one, and it resonates today: high technology is good, but lower technology as part of a well-organized system can be better. CH radars were one part of the CH radar system. Behind the radars there was an integrated network of CH command and control equipment and personnel that made all the difference. I'll tell their story in this chapter, but first I will show you how the CH radars worked (more or less).

CH Radars

In contrast to all successor early warning radars, CH applied the floodlight approach. That is to say, the skies to the south and east of England were

---

[12]Nevertheless, modified and improved CH radars continued to operate as Britain's air defense long after the Battle of Britain in that warm summer of 1940. The CH system continued to the end of the war. In 1944 it was adapted to detect incoming V2 rockets destined for London.

filled with HF radio waves. This is because the CH transmitter beams were not steerable—they were far too big for that—and so to provide coverage without gaps it was necessary to flood the skies. It is much more efficient to search a large volume of air by scanning a narrow beam rapidly across the sky, which is what we do nowadays with air defense radar systems, but this option was not open to the CH designers.

The transmitter antenna of each CH station consisted of four large steel girder towers, with cables (which emitted the radio waves) strung horizontally between them. The towers were 110 m (360 feet) high and 55 m (180 feet) apart[13] (fig. 1.1). The average height of the cables was 65 m (215 feet). Viewed end-on these cables would look like the vertical linear array of figure T4.3, and so you can see that the CH transmitter assembly formed a beam. The mainlobe was about 100° wide in azimuth—to provide wide coverage—but with much narrower elevation coverage, from 2° above horizontal to 16°. This elevation coverage was quite adequate to detect aircraft over France and approaching the English coast. It would be a waste of power to direct radio waves to other elevations if detection was not needed there; hence the restricted elevation extent. This system had a gap, however, at about 5°, which needed to be plugged, or enemy planes at this elevation would not be observed. So an auxiliary array was placed on the towers at height 30 m (95 feet), which filled the hole.

Receiver antennas were placed some distance from the transmitter array to avoid interference effects.[14] The receiver towers were wooden, 70 m (240 feet) high, with array antennas—crossed dipoles, for the cognoscenti— at the same heights as the transmitter antennas (mainlobe and gap filler).

These transmit-and-receive towers constituted the main hardware of one CH station. At the time of the Battle of Britain, the transmit frequency was 25 MHz ($\lambda = 12\,m$), though this frequency was increased (wavelength shortened) later on in the war. Transmit power was 350 kW, later increased to 750 kW. The power source was water cooled. The HF radio waves were transmitted in pulses of 20-μs duration, which provides a pulse length of 6 km and hence a rather coarse range resolution of 3 km. Why so coarse? After all, it meant that two enemy airplanes, one following within 3 km of another, would be mistaken for one plane. Again the answer has to do with

---

[13]Several such towers still stand. One is at Great Baddow, in Essex, on the site of an aerospace company that still designs and builds electronic warfare equipment.

[14]Such as "cross talk" or other electronic gremlins, which I will not discuss further. These interference gremlins were known as "running rabbits" to the CH engineers.

Figure 1.1 *Surviving chain home transmitter towers. The tower on the right in the top photograph has modern antennas attached. Thanks to Barry Stewart for permission to reproduce this photo. Sunset image is from Wikipedia.*

detection range. Shorter pulses contain less energy and so can detect a target at shorter range, as we will see in the next chapter. The effective range of the CH radars was 150 km (90 miles), which gave about 20 minutes warning.

The time between pulses (*pulse repetition interval,* or PRI) of 40 ms (milliseconds) was chosen carefully, so that one CH station would not get confused by the echoes from another station. Let me explain. CH stations were

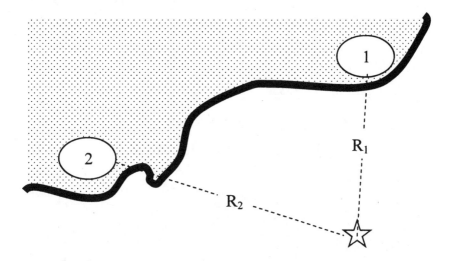

Figure 1.2 *Two Chain Home radar stations (ovals) on the south coast of England (bold line). They both detect an incoming airplane target (star). To avoid interfering with each other, the signals transmitted from stations 1 and 2 must be carefully coordinated.*

sited carefully so that the transmitters and receivers did not interfere—for example, with a hill between stations. However, the coverage of adjacent stations must overlap, so that there was no gap in the curtain that cloaked the English coast. But overlapping coverage meant that the transmitters from two different CH stations (fig. 1.2) might be illuminating a German plane. The transmitted pulses would reflect off the plane, and back into the receivers of both stations. So how would one station know which of the echoes came from its own transmitter, and which from its neighbor? It was important to know the pulse origin, since ranging was achieved by careful timing of each pulse, as we have seen in note T2. Mistaking your neighbor's pulse for one of your own would lead you to incorrectly estimate the airplane range. To get around this problem, the transmitters all over England were synchronized to the National Grid (the national electricity supply), so that the pulses from different stations could be staggered (fig. 1.3).

Consider figures 1.2 and 1.3. Station 1 transmits a pulse that is reflected back into its receiver from the target. But the station 1 pulse is also reflected back into the station 2 receiver, and vice versa. So if stations 1 and 2 both transmitted their pulses at the same time, then the target would appear as

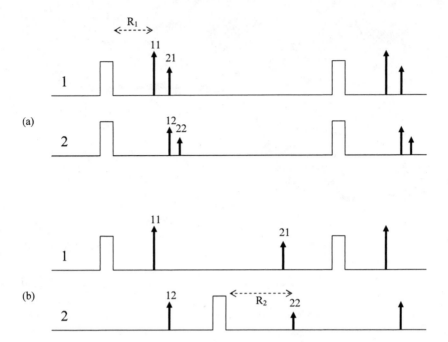

Figure 1.3 *(a) Stations 1 and 2 emit synchronized pulses. The echoes are labeled as follows: 12 means the pulse from station 1 received at station 2, and so on. The time delay from pulse to echo depends on range and is shown for echo 11. The single target of figure 1.2 has generated more than one echo in both receivers. The spurious echoes represent misinformation resulting from interference. (b) The pulses from stations 1 and 2 are staggered in time. Now the spurious echoes are restricted to the latter half of the pulse repetition interval and can be rejected. True targets will appear only in the first half.*

shown in figure 1.3a. At station 1 the true target signal is detected at the correct range R_1 (recall from note T2 that delay time is proportional to range), as shown. But there also appears to be an echo from range $\frac{1}{2}(R_1 + R_2)$, from the pulse that originated at station 2. Similarly a double echo appears in the station 2 receiver, as shown. These spurious echoes are dangerously misleading, and must be eliminated. The CH system achieved this by staggering the pulses from different stations, as shown in figure 1.3b. Now station 2 emits each pulse 20 ms after station 1. If you stare at figures 1.2 and 1.3 for a short while, you will see why this eliminates the problem. The true signals appear in the right place: the target shows up in the station 1 receiver at range R_1 and in the station 2 receiver at range R_2 (shown). The spurious echoes now turn up in the trailing half of the interpulse period, in both stations 1 and 2. So, by ignoring anything that appears in the second half, each CH station eliminates the echoes that originated from the other station.

In reality more than two stations might see a target airplane at any given instant, so that the interpulse period had to be divided up into more than two sectors. Synchronization was sufficiently accurate to permit this division to be achieved reliably across all the CH transmitters. This is the main reason why the PRI was so long. Forty milliseconds corresponds to a 6,000-km range—far greater than the detection range of the radar system— but to defeat the spurious echo problem this 40 ms had to be divided into many nonoverlapping time slots.

So the range of an incoming plane could be estimated unambiguously by the CH system, but what about direction? The bearing (azimuth angle) was found by the technique of *amplitude monopulse*, still in use today. Recall that, for coverage, the CH transmitters had a very broad beam (100°) and so were no use at all in providing information about target bearing. The receiver antennas were narrower, however, though not narrow enough to provide a good estimate. Suppose, for example, that a target is detected at $R = 100$ km range, by a beam that is $\Delta\theta = 20°$ wide: the width of this beam at the target range is $R\Delta\theta = 35$ km. This width is too imprecise a location estimate to direct intercepting fighters—they would likely just find a bunch of air. Similarly, at long range the elevation beamwidth corresponded to a considerable span of target airplane altitudes. Amplitude monopulse fixed this also, as explained in note T6.

The estimate of target altitude was a little more complicated than that for target bearing, because of the presence of ground reflections. Transmitted power would reflect off the ground in front of the station, and upward, to change the effective shape of the elevation beam. This fact, of itself, was not a show stopper, so long as the CH designers knew that the power was indeed being reflected and was not absorbed by the ground. This is another reason why the transmitter sites were chosen with great care.[15]

---

[15]There were five reasons why the siting of CH stations required careful consideration. I have mentioned three of these already: the transmitters from one station must not interfere with the receivers of the same station; the transmitters of different stations must form an unbroken floodlight curtain with no gaps; and the land in front of the transmitters must be flat to provide constant known ground reflection power. The fourth reason is logistical: the sites, which were often remote, must be convenient for building on and supplying, and must be secure from offshore bombardment. The fifth reason, stated in reports at the time, is that the sites could not "gravely interfere with grouse shooting." I kid you not.

Ideally, the land in front of the antenna should be flat; this was rarely the case, which is why the CH towers were located on the coast. The flat sea reflected like a mirror.[16] The CH system, I should note, was thus outward looking. It covered the English Channel and the southern portion of the North Sea that was adjacent to Eastern England. Once enemy aircraft got past the coastal chain of CH stations, and were over land, the CH system was useless. When this happened, observers on the ground determined enemy aircraft location. However, when the system worked properly the invaders would be met by a welcome-wagon of RAF fighters before they reached land, as we will see, so that the enemy was thereafter always in sight. Another slight complication with elevation estimate was in converting target elevation angle into target altitude. At long range, this requires consideration of the earth's curvature. The CH designers knew about this complication and took it into account.

The elevation coverage was poor below 2°, and this low elevation blindness (deafness, using the analogy of the introduction) was a weakness of the CH radars. Enemy aircraft could sneak in under the radar, by flying low, close to the sea surface. This fact became known to the Germans in time, but was not widely exploited during the Battle of Britain. Later in the war the leak was plugged, as it were, by the introduction of CHL (Chain Home Low) which could detect targets at lower elevations. In chapter 4 I will explain why radars have trouble at low elevations, and how they overcome these problems. For now, though, please just accept that the CH stations in 1940 were, metaphorically speaking, unable to hear the roar of low-flying enemy aircraft over the hiss of the sea.

The CH System

At the outbreak of World War II, British radar development was not as advanced as that of the United States. More importantly at the time, it was not as advanced as German radar development. But (and this is the crucial factor) it was only in Britain that the tactical use of radar, in a real battlefield situation, had been thought through. German scientists and engineers (and their U.S. counterparts) were concentrating on technological advances.

---

[16]Here is one advantage of the long-wavelength HF radio waves, compared with shorter microwaves. In rough seas the microwaves scatter at all angles, but radio waves do not notice such small-scale variations in the reflecting surface. So, HF transmitter power that is reflected off the sea does not vary with sea state, and the effective shape of the CH elevation beam is constant— a boon to target altitude estimation.

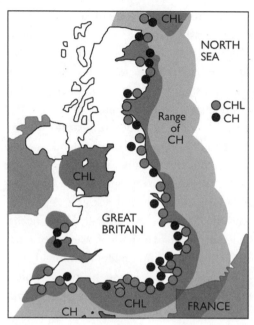

a

Figure 1.4 (a) Chain home (CH) and chain home low radar (CHL) sites and ranges. Note that northern France, where many of the Luftwaffe bases were located, was within range. (b) Distribution of Luftwaffe air fleets. Images adapted from maps made available to the author by the Battle of Britain Historical Society.

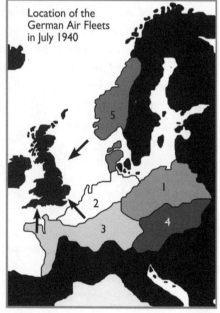

b

In Germany the military authorities were ambivalent about radar; they liked technology but were less than enthusiastic about a system they (wrongly) regarded as purely defensive. German scientists were good at technical research and development, but military tactics were not part of their ambit. Tactics and strategy are the heart and soul of military men, of course, but in Germany these men did not understand or fully appreciate radar.

In 1939 Britain, there was no such fault line between radar technology and radar deployment.[17] For this happy and unusual circumstance we can thank Air Marshall Hugh Dowding, a prickly and rather remote man, whose nickname at school (Stuffy) may not be wholly inappropriate. Dowding was, for most of the 1930s, responsible for RAF supply and research. An enthusiastic convert to radar following the Daventry experiment, Dowding listened while Watson-Watt explained his vision of a chain of interlinked radar stations. Funding followed. In 1936 the RAF set up Fighter Command, with Dowding in charge. The air defense of Britain was now his responsibility, and he set about integrating the rapidly developing CH radar system with his fighter aircraft squadrons.

It was realized quite early that the information supplied by CH radars could be very confusing and misleading. We have seen how the overlapping coverage of several CH stations (a military necessity) might have led to spurious target sightings and to incorrect range estimation. Imagine, then, the potential confusion caused by 20 stations reporting their findings, along with the Observer Corps filing reports on enemy locations and numbers, and the radio reports from airborne fighters. One errant bomber, off course, might generate dozens of sightings by eye and by radar; and to Fighter Command this might appear as a major assault by many enemy aircraft. To avoid such errors and to make sense of the anticipated information flow, Dowding's team installed a "Filter Room" at Fighter Command

---

[17]This fault line between technological development and military application is more than likely the norm rather than the exception. Certainly it existed in the United States during the early war years, as we will see, and continues worldwide to this day. I can testify from personal experience, over two decades' employment with large and inefficient British-based aerospace corporations, that such bungling is alive and well, gold-plated and surrounded by neon lights, with trumpet fanfare and corporate advertising, and with a budget dwarfing that available to Dowding and Watson-Watt. Perhaps I am being a little unfair here, since new technology does not always work, and it is not always possible to foresee the best way to apply it when it does work. However, I am sure that many readers who have experienced government-funded research programs will recognize something familiar in the situation I describe. Weep, then, tears of joy at the harmony achieved in Britain in the late 1930s between radar R&D and radar application.

Headquarters (HQ), manned (and womaned) by experienced personnel to sift the data quickly, before passing it on to the General Situation Map. This map would display the best estimates of enemy numbers, location, and disposition to the Air Marshall and his subordinates. In addition to this main map, there were table maps in the Operations Room. These maps were divided into grid squares, and colored blocks representing groups of British and German aircraft were moved over the map, as in a board game. This image is a movie cliché today, with the blocks moved croupier-style by nubile girls in uniform, while gray-haired brass hats look on with furrowed brows. In fact, most of the Operations Room staff were indeed young women in WAAF (Women's Auxiliary Air Force) uniform.

From the Operations Room, several Controllers would telephone the situation, and operational orders, to the various fighter air bases. Further orders would then be transmitted to the pilots by radio, once they had scrambled their planes and were heading toward the anticipated combat zone. A lot of effort went into developing communications in the burgeoning infrastructure of the CH air defense system: communications between CH radar stations, the Observer Corps, and the Filter Room at HQ, between Fighter Command and the airfields and airborne fighters. Much thought also went into what we today call *ergonomics*, the study of the efficiency of workers and of working arrangements: how best to display information on the maps and how best to relay detailed information quickly to the airfields and to pilots.

The rapidly developing system was tested as far as possible, to find any weak links. Was the radar coverage complete? Were inbound hostiles correctly located? What signal *bandwidth* should the CH operators use?[18] (The concept and relevance of bandwidth is explained in note T7). Was the information passed on quickly and accurately to the airfields? How quickly could a fighter squadron scramble into the air? These tests uncovered the CH shortcomings at low elevation, and before the Battle of Britain began there were moves afoot to install the CHL fix, but this was not ready in time. Indeed, time was very pressing, as I indicated earlier. One source claims that Dowding recommended to the Prime Minister, Neville Chamberlain, that he appease Hitler at Munich to gain time. The implication here is that the sad spectacle of appeasement, universally interpreted

---

[18]There were three operating bandwidths available for the CH system: 50 kHz, 200 kHz, and 500 kHz. The wide bandwidth was used for raid-strength assessment, while the narrow bandwidth was more suitable to combat enemy jamming.

then and now as weakness in front of the bullying and aggressive Führer, in fact served a better purpose. But other historical accounts are silent on this matter.

The Germans were naturally suspicious of the large masts that appeared around the coasts of southern and eastern England, facing the approaches from France and Scandinavia that would be likely invasion routes. In 1939 they sent a Zeppelin to cruise over the English Channel and gather information. The Zeppelin crew photographed the towers and intercepted the EM transmissions. They concluded that the towers were not radar transmitters, but were instead intended for radio communications and naval rescue. This error of judgment was a significant blunder, with major consequences to follow. A German source—in a position to know—claims that the Zeppelin looked for higher frequencies emanating from the CH transmitters and that, finding nothing above HF, concluded that the towers could not be radar transmitters. The frequency was too low, and surely the British had moved on from the primitive radio detection devices of the previous decade? If true, this fortunate (from the Allies' perspective) misidentification is a result of Watson-Watt's "third best" policy—choosing low-tech HF rather than state-of-the-art microwave operating frequency.

The CH system was primitive and was rushed into service. Scientists and engineers worked desperately to design and develop, manufacture, install, and test the equipment in the years, months, and weeks before the invasion. After the German *Blitzkrieg* against France, Holland, Belgium, Luxembourg, Denmark, and Norway (lasting only a few weeks), everybody knew that England was next in line. In June 1940 Churchill—by now he was Prime Minister—said "the Battle of France is over. I expect the Battle of Britain is about to begin." The CH infrastructure was still being developed when the German air raids began, and CH operators learned on the job as much as they learned from training exercises. There were glitches and errors, and the CH system technology relied on nothing new, but in the summer of 1940 it was the best (indeed the only) complete air defense system in the world.

Kanalkampf

Just after lunch on July 10, 1940, a convoy of merchant ships heaved abreast of Dover, a large sea port on the southeast coast of England. The CH radar system indicated to Keith Park, RAF commander-in-chief of 11-Group, covering the Southeast sector of England, that enemy aircraft were massing behind Calais, the French port opposite Dover. Park cautiously sent six fighters, Hurricanes from No. 32 squadron—already airborne and in the area—to intercept them. The Hurricanes found 70 German planes (fighters and bombers) attacking the convoy, and immediately pitched into them. By the time that more RAF cavalry arrived, in the form of 20 Spitfires and Hurricanes, the Germans had adopted a more defensive formation, with bombers circling low over the convoy, while their fighter escorts were higher up, fending off the six Hurricanes. A dogfight ensued: four *Luftwaffe* (German Air Force) planes and three RAF fighters were downed. One convoy vessel was sunk.

This initial action, small though it may have been, was typical of this first phase of the Battle of Britain (*Kanalkampf*—"Channel war"), and brought to the fore a dilemma facing the RAF. Before describing this dilemma, and the course taken by the battle, I need to discuss numbers: the number of planes ranged against each other during the battle (and this is the only battle in history that was fought entirely in the air) and the number of planes shot down. Both sets of numbers are disputed, and different sources quote different numbers.

When addressing the number of aircraft available for each side, we should really consider only those that were serviceable at the time, and which were available for operations over Britain. British bombers will not be included since, with one exception, they took no part in the battle. Bombers are offensive weapons and so were of no use to the RAF fighting a defensive battle. The Germans were very much on the offensive, so I will include bombers in their roster. So, there were about 650 fighter planes in the RAF that were ready and available for action in July 1940. Losses throughout the battle were made good by an efficient Minister for Aircraft Production, Lord Beaverbrook.[19] These planes were mostly Hurricanes, the workhorses, with a little over half their number of Spitfires, the thoroughbreds. Both types

---

[19]Beaverbrook was born a Canadian, and Keith Park, who would play a crucial role, was a New Zealander. The non-British contingent involved on the Allied side during the Battle of Britain was significant, and this is a good place to enumerate them. The official figures for RAF participants (pilots and ground staff) consist of 2,340 British, 145 Poles, 127 New Zealanders, 112 Canadians, 89 Czechs, 32 Australians, 28 Belgians, 25 South Africans, 13 French, 10 Irish, 9 Americans, and 6 others. Of these, 544 would be killed during the battle. Many of those from overseas (such as the Americans) were volunteers and many played a key role, at a time when experienced pilots were at a premium. The Poles and Czechs, for example, were air force pilots who were forced to flee their own countries after Nazi occupation, and who chose to carry on their fight in English skies.

Figure 1.5 (a) A World War II
photo of flying Spitfires. (b) Eagle
in a Hurry: the insignia of the
U.S. volunteer Eagle Squadron emblazoned on a Hurricane. Thanks
to Stephen Wilkinson for this picture. (c) Spit'n'Polish: a Spitfire
of one of the Polish squadrons.
Thanks again to Stephen
Wilkinson.

a

b

c

were largely unknown quantities, because they were relatively new designs[20] (1936–1937—brought in under the aegis of Dowding, who recognized the need for modern fighters). German planes amounted to about 2,500 serviceable aircraft, fighters and bombers. Their losses during the battle were replaced only slowly—they were not used to taking losses. The fighters were mostly Messerschmitt Bf 109s, generally considered to be the best fighters in the world at that time. The bombers were small by later standards, mostly two-engined, but they included the fearsome Stuka dive bomber that had played such a prominent role during the fall of France, and earlier during the Polish campaign.

Loss assessment is difficult because both sides tended to overestimate the number of enemy planes shot down. This was partly propaganda, no doubt, but there was also a more legitimate reason. In the heat of battle several pilots may engage an enemy plane, which they then see spiral out of the battle, trailing smoke. More than one of these pilots may claim the kill, and so one downed plane becomes many. Also, the plane may not have been destroyed; it may have staggered back home or successfully landed on friendly soil—a significant home advantage for the British. Perhaps because they could count downed planes (at least those on land) the RAF tended to make less exaggerated claims of enemy losses than did the Luftwaffe; even so, they often claimed three times the number of kills achieved on a given day. The median numbers that I find from several historical sources suggest that the RAF lost about 800 fighters during the three months of the Battle of Britain, while the Luftwaffe lost about 1,500 fighters and bombers.

So much for numbers. The first phase of the battle consisted of a Luftwaffe offensive reconnaissance, aimed at Allied convoys approaching English ports, and at radar stations and towns on the south coast of England. The dilemma faced by the RAF was this: how much do they commit to action over the sea, and how many planes do they send against an enemy formation? Dowding argued against sending fighter planes to cover the British retreat from Dunkirk earlier the same year. He also chose to fight over England, rather than engage the German bombers over France or far out

---

[20]Both proved to be successful planes, judging by the numbers that were produced during the war. The Spitfire performed better, matching the best German fighters. It was designed by Reginald Mitchell, while he was dying of abdominal cancer. He did not live to see his Spitfires in action. At one point during the battle, the Luftwaffe C-in-C (Reich Marshal Goering) asked German ace pilot Adolph Galland what he needed to help him in battle. "A squadron of Spitfires," Galland replied.

in the Channel, as they approached the English coast. The idea was to pre-
serve fighters, since the disparity in numbers between the two sides dictated
caution and economy. However, there was a counterargument (engage the
enemy before they can do harm) and Dowding's decision was controversial
at the time.

Park's initial reaction accorded with Dowding's view. He scrambled his
planes in small numbers, based on CH radar system estimates of enemy
strength and location. This cautious approach limited his losses—he feared
he might lose the battle on day one if he sent out all his planes—but of
course it also limited the damage inflicted. Other RAF commanders, principally Leigh-Mallory of 12-Group covering the English midlands, thought
differently; they wanted large formations to engage the enemy and knock
them out of the sky. The disagreement grew throughout the battle, and
beyond.

This initial phase of the battle lasted four weeks. During this time the
Luftwaffe probed the defenses, softening up the invasion routes, and, most
significantly for our story, bombing the CH radar sites. This bombing was
half-hearted, because the significance of the towers was not fully appreciated, as we have seen. Also the German dive-bombers, brought to bear so
successfully in earlier campaigns in Europe, did not like the CH towers.
The antenna cables slung between the towers were an effective barrier
against low-flying aircraft. The towers were not so easy to knock over, since
they were constructed of steel girders; consequently, there was not much to
show for the bombers' efforts. Even so, this period was a critical one for the
defenders. The Germans were further discouraged when they observed that
a bombed tower would seem to continue transmitting. In fact, this was a
bluff. Sometimes the CH stations were damaged, and radar coverage com-
promised, but fake transmissions (not part of the CH system) were sent out
to imply otherwise. A concerted Luftwaffe effort could have knocked out
the CH system, but this did not occur.

During the Kanalkampf phase, the RAF lost about 150 planes, and the
Luftwaffe about 250. This may look like a victory for the home team but,
given the disparity in numbers, it was not. Dowding was well aware of this
fact—in particular, of the loss of experienced pilots. The Germans had also
sunk a large amount of allied shipping, so that the Admiralty was obliged
to reroute convoys away from the Channel. On the positive side, the CH
system worked. It would come into its own during the crucial second phase
of the battle.

Adlertag

Hitler's Directive No. 17 ordered that "The Luftwaffe is to overpower the Royal Air Force in the shortest possible time." Goering figured it could be done in four weeks. The operation for the invasion of England, code-named *Seelöwe* (Sea lion) required dominance in the air, because otherwise the RAF would cause havoc among the invasion fleet. Hence, Directive 17. The destruction of the RAF was scheduled originally for August 8, but *Adlertag* (Eagle Day) was postponed until the twelfth because of bad weather. Eagle Day launched the main offensive of the battle, and this phase lasted until the first week of September. The Luftwaffe sought to destroy the RAF in the air and on the ground. Eagle Day was preceded by an attack on coastal CH stations, and succeeded in disabling three of them, though they were up and running again after six hours.

On Eagle Day the Germans believed that they had downed 84 RAF fighters and destroyed 8 air bases. This was largely illusory; the claimed kills were wildly overestimated and new planes were delivered promptly to replace actual losses. Damaged airfields were repaired sufficiently to keep operating. The pattern was set for the next several weeks. CH would detect Luftwaffe formations approaching the English coast, and just enough fighters would be scrambled to meet them in the air. The Spitfires and Hurricanes concentrated on the bombers (Dowding's priority). The Messerschmitt fighters had limited range and could manage only 20 minutes' flying time over Southern England. Dogfights filled the sky, and BBC commentators reported them in the manner of live sports events. "You can't see these fighters very coherently for long. You just see about four twirling machines and you hear little bursts of machine gun fire, and by the time you've picked up the machine they've gone . . . There's a dogfight going on up there—there are four, five, six machines, wheeling and turning round. Hark at the machine guns. Hark one, two, three, four, five, six. Now there's something coming right down on the tail of another. Here they go—yes, they're being chased home and how they're being chased home. There are three Spitfires chasing three Messerschmitt's now . . . Oh boy! Look at them going. And look how the Messerschmitt's—oh that is really grand! And there's a Spitfire just behind the first two—he'll get them!"

The CH system was under strain, but the value of it was becoming clear; in modern military parlance the radar advantage was a force multiplier. The outnumbered RAF need not waste time and effort with standing patrols, hoping to encounter incoming "bandits" by chance. Instead, they knew ahead of time where the bandits would be, and often had some idea how many of them would be there. So they sent up just enough fighters (angels), at just the right time, to intercept and disperse them. Some Luftwaffe low-level raids got through, passing over damaged CH stations that were temporarily out of service, and these raids achieved complete surprise and did a lot of damage to airfields. Sometimes two raids would be mounted at the same time, arriving at different altitudes and aiming for different targets, to stretch the defense.

One very large raid on the North of England occurred on August 15. This raiding party originated from German airfields in occupied Denmark and Norway, and was severely mauled. (The Germans had insufficient fighter protection, and believed that almost all the RAF fighters had been sucked into the battle further south.) Other raids took place in bad weather, or at night. These were difficult to intercept. Radar would spot the invading formations but the fighters sent up to intercept them could not pick out the raiders by sight. CH accuracy was not sufficient to pinpoint enemy planes, so the final approach had to be made by sight. Henry Tizard, chairman of the Aeronautical Research Committee, realized quite early what was needed—airborne interceptor (AI) radars. These were being developed during this phase of the battle, and prototype versions would be employed successfully later.

Luftwaffe tactics were adjusted. The Stukas were taken out of the battle—they may have been devastating against ground targets during earlier campaigns but now they were vulnerable to Hurricanes and Spitfires, and too many Stukas were being shot down. Goering ordered that attacks on radar bases be halted, so as to concentrate more on the airfields. Big mistake, but a bigger one was to follow.

Often the damage inflicted on an airfield, say Biggin Hill in Kent, would be fixed during the night, so that when the Luftwaffe planes returned next day they still faced fighters operating out of Biggin Hill. To deter these repairs the German bombers learned to return on the same night. The difficulty of locating enemy aircraft in the dark worked both ways, however. On one fateful occasion, on August 25, a Heinkel He 111 bomber got lost and accidentally dropped its bombs on the city of London—off limits previously, on Hitler's orders. Churchill ordered a retaliatory raid on Berlin. The following night, 80 RAF Hampden bombers carried out his order. Hitler was incensed, and ordered the eradication of London. From September 7, the people of London faced the Blitz, and this switch of targets by the Luftwaffe saved the RAF.

The Blitz

The consensus of historians today considers that the RAF was within a week or two of collapse, when the bombing of London began. The CH system was stretched, airfields were repeatedly raided and patched together, and fighter pilots were exhausted. The problem of acute pilot fatigue is a consequence of the force multiplier effect of the CH radar system. Instead of one day of action and several days of patrols, pilots were faced with many actions in one day,[21] and were "on call" any time. Nuisance raids by single planes, designed purposefully to alert the British defenses and dissipate resources, only exacerbated this problem. But the change of Luftwaffe targets changed everything.

First, because they were no longer in the firing line, the airfields, CH stations, and communications infrastructure could be repaired, and pilots rested. Second, the Messerschmitt fighters had less time over London—10 minutes only—than over the more southerly airfields of Kent and the southern coast. Consequently, the bombers were less protected, in particular, during their daylight raids. Third, the longer flight time to London meant that CH gave Fighter Command more notice of an impending attack, so that there was more time to get planes in the air. Also, the skies above London were close to Leigh-Mallory's 12-Group, as well as Park's 11-Group, and there was sufficient time to form the large formations of Spitfires and Hurricanes that Leigh-Mallory favored—a luxury not available earlier.

The Luftwaffe formations that attacked London had more fighters and fewer bombers than before. The intention here was to coax up the last few RAF fighters (so the Germans thought) to their destruction. Late in the morning of Sunday, September 15 (nowadays celebrated as "Battle of Britain Day"), a large formation of more than 1,000 Luftwaffe planes approached the East Kent coast, heading for London. The crews were astonished to be greeted by more Hurricanes and Spitfires than they thought existed. One "big wing" of 60 planes included British, Canadian, Czech, and Polish squadrons. Spitfires tackled the Luftwaffe fighters, while the Hurricanes went for the bombers. It so happened that the Prime Minister was visiting Park at 11-Group HQ that day, and saw the combat build up. The bombers

---

[21]There are several anecdotal stories of RAF pilots who were shot down, landing unharmed, or who bailed out of burning planes, catching taxis back to their airfields, and going back into action later the same day.

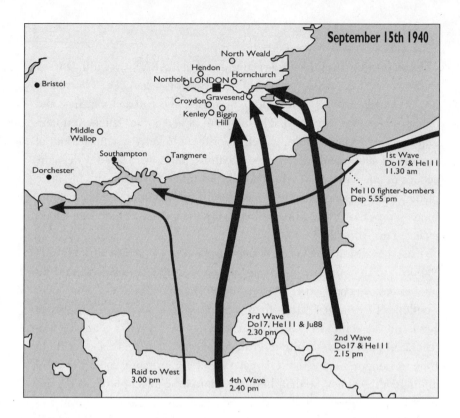

Figure 1.6 *Luftwaffe raids at the height of the Battle of Britain. Image adapted from a map made available by the Battle of Britain Historical Society.*

never reached their targets; they scattered and headed for the coast. That afternoon a second large raid was attempted, with the same result. The Luftwaffe lost 56 aircraft that day (the RAF thought it had shot down 185). Two days later Operation Sea Lion was postponed, and a month later it was canceled.

More large raids followed, but increasingly they happened at night, to reduce casualties, and were directed at other British cities: Birmingham, Southampton, Bristol, Liverpool, Coventry, Sheffield, Glasgow, and Belfast. They were no longer intended to destroy the RAF, but rather to destroy British industry. The RAF weapons opposing these raids were now specialized night fighters (two-engined Beaufighters equipped with AI radar) rather than Spitfires and Hurricanes. The nature of the battle had changed. It only gradually became clear that the Germans had conceded defeat.

The London Blitz lasted from September 7, 1940, until the following

Figure 1.7 *The London Blitz. "Standing up gloriously out of the flames and smoke of surrounding buildings, St. Paul's Cathedral is pictured during the great fire raid of Sunday December 29th," 1940. National Archives (photo no. 306-NT-3173V).*

May. It was then halted because Hitler was preparing his massive offensive against Russia, so troops and materiel were transferred east. During the Blitz more than 40,000 civilians were killed—the era of strategic bombing had begun. Again, ascertaining numbers is problematical, as they vary with the source consulted. Figures for civilian casualties start at 46,000, and perhaps 1,000,000 homes were destroyed or damaged. (These numbers would later be dwarfed by the casualties resulting from Allied strategic bombing of German cities, but at the time both sides considered the London Blitz to be the most devastating aerial assault on a city that the world had yet seen.)

The outcome of the Battle of Britain is significant for two reasons. First, it raised morale of those peoples around the world who opposed the Nazis, and second, it kept Britain in the war. Morale was raised because the Battle of Britain was the first defeat of the seemingly invincible German war machine. In America, public opinion swung round to the view that Britain could, with help, win the war. This brought the United States closer to war with Germany. The fact that Britain stayed in the war[22] proved to be of great importance later on. Russia was provided with crucial supplies (American and British) from British ports. In spring 1944 the southern coast of England—the very region where most of the Battle of Britain was fought— turned into a massive staging post for Allied troops and materiel. The D-day landings could not have been launched from anywhere else.

So what were the factors that tilted the balance, leading to this first German setback, despite the odds? Well, home advantage certainly helped, and the RAF made fewer tactical blunders[23] than did the Luftwaffe. Also, we now know, Dowding benefited from Ultra, the code-breaking system that enabled Churchill and his senior military staff to read Hitler's mail. In particular, Dowding knew that September 15 was make-or-break day for the Luftwaffe; Goering had been told that the invasion would have to be postponed if he could not break the RAF by mid-September, because bad weather would preclude landing operations until the next spring (though tides, moon, and weather all favored a large-scale amphibious landing on the fifteenth, and for a few days afterward). So, the massive attack on that morning was anticipated. When it failed, Goering turned his planes around and sent them back again—a desperate last throw of the dice. Dowding knew that this was the day when every available fighter should be up in the air, and this is the reason why Luftwaffe crews were so astonished by the number of Spitfires and Hurricanes that opposed them. It now seems that Churchill did not just "happen to be" at 11-Group HQ that morning— though all accounts of his movements reported it that way, then and for the next 35 years.

But good intelligence does not win dogfights. The force multiplier effect of the CH radar system was absolutely essential. It placed the RAF fighters

---

[22]For a year—June 1940 until July 1941—Britain was the only major power opposing Hitler. America was neutral, France was beaten, and Russia had a nonaggression pact with Germany. Britain was actively supported by former "Empire" countries such as Canada, New Zealand, Australia, and South Africa, but these were very distant.

[23]The disagreement between Park and Leigh-Mallory over tactics carried on after the battle. The Air Ministry supported Leigh-Mallory, who later took over Dowding's job at Fighter Command. Dowding was retired a few weeks after the battle; his tactics had been called into question, and he had stepped on too many toes during his tenure at Fighter Command. The key contributions made by both Park and Dowding would later be recognized by their country, however, and historians are in no doubt: Dowding set the foundations for victory before the war began, and Park effectively managed his crucial sector while under heavy attack.

where they should be, and at the right altitude, to meet their enemy. Ultra brought great strategic benefits to the British then, and to the Allies later, but Ultra was useful only for large-scale strategic issues. It was too big a secret to be applied to tactical matters, no matter how important[24]—unlike radar. The CH system identified groups of incoming "bandits" whatever their numbers, whatever the significance or insignificance of their mission. Radar had shown itself to be a *tactical* weapon of supreme importance, when welded to a command and control structure that believed in it. In time the power that radar conferred on armed forces became clear to all the nations that fought during World War II. The breathtaking speed with which radar technology evolved during the war is testimony of its importance. Almost all the radar signal processing techniques that we use today can be traced back to these early years following the summer of 1940, when radar first burst on the scene of battle.

From the Tizard Technical Mission to Pearl Harbor

The Battle of Britain has served its purpose here, to introduce basic radar concepts, and I do not intend to describe in detail the whole of World War II. I cannot close this chapter, however, without mentioning some of the more important technical developments that arose later on in the war.

At the height of the Battle of Britain, Henry Tizard led a technical mission across the Atlantic to the United States. This mission included many of the leading scientists and engineers who had developed British radar, and other technical innovations of military significance, such as jet engines and rocketry. They arrived with blueprints, operating manuals, and some key components of working equipment. Tizard and his team were welcomed by eminent American scientists and military men under the chairmanship of Dr. Vannevar Bush. The U.S. team included Karl Compton, a well-

---

[24]The bombing of Coventry provides a tragic example of the importance attached to Ultra. During the Blitz phase of the battle, Churchill and his advisers were provided with advance knowledge of a large-scale Luftwaffe raid on this industrial city, located in the English midlands. They knew that operation "Moonlight Sonata" would take place on the night of November 14/15, 1940. Churchill decided that nothing should be done to warn the city and so reduce civilian casualties. No extra defenses were allocated to Coventry, no extra ambulances were sent there; no tip-off was given to the emergency services. It was considered that German intelligence may infer (correctly) that Hitler's top-level cipher code "Enigma" had been broken, had any unusual activity been noted in Coventry ahead of the raid. Subsequent events showed that the Ultra secret was worth keeping, but the people of Coventry paid the price. Five hundred and fifty-four civilians were killed that night, 865 were seriously wounded, and the city lay in ruins.

known physicist and president of the Massachusetts Institute of Technology, Admiral Howard Bowen, director of the NRL, along with the president of Harvard University and others. Why would the British send, at some risk and at a key moment of crisis at home, their most eminent scientists and engineers to a neutral country? To give away their nation's secrets.

I first learned of the Tizard mission 20 years ago, at the beginning of my career in radar systems analysis. It amazed me then, and amazes me now, to think that a belligerent country could entrust its most cherished technological secrets, of direct and immediate military significance, to a neutral state (more than a year before America joined the Allies). Yet such was the case. The British realized that they could not win the war without the enormous research and productivity capacity of the United States. So, in exchange for technological goodies, they hoped that America would gear up to supply Britain with weapons of war. The decision had been made after the fall of France in June 1940. In July Churchill met with President Roosevelt and the mission date was set for September.

Tizard's team talked, and the Americans listened. Here I will concentrate on the radar matters that were discussed. The British revealed their ASV Mk 2 airborne ship-detection radar, and their AI Mk 4 night fighter radar. Both of these had been tested in war, and both were immediately installed in U.S. aircraft and tested further by U.S. military engineers and scientists. The United States, in turn, demonstrated its own radar advances, and it became clear that American metric radars (those with operating wavelength of about 1 m) were just as advanced as the British. Importantly, however, the U.S. equipment was not integrated into a coherent radar system and it had not been tested in battle.

The most important innovation disclosed by Tizard, without doubt, was the cavity magnetron developed by J. T. Randall and H. A. H. Boot of Birmingham University. This device is a source of microwave radiation, and was to become the heart of later microwave radar transmitters. The basic idea was not new—both Germany and the United States could lay claim to earlier magnetron inventions—but the Randall and Boot multicavity version was much improved: at about 50 kW it produced *100 times* the power of earlier variants. This made microwave radars (in this case, with a wavelength of about 10 cm) feasible. Such short wavelengths led to improved angular accuracy, as we have seen, and this accuracy was vital for detecting small targets. Also, the transmitter was small, and so could readily be installed on aircraft. It is not difficult to imagine jaws dropping on tables when the multicavity magnetron was unveiled.

The Americans immediately recognized the importance of this device. The MIT Radiation lab was set up, under Lee DuBridge, to develop microwave radars, largely on the strength of the Tizard mission disclosure. No less than half the radar designs developed worldwide during World War II would come out of this lab. The United States would produce over one million cavity magnetrons during the war.[25] Up until this time, British and U.S. radar development had been separate, but the Tizard mission brought them together, and it was decided to split up further radar R&D. The U.S. share was microwave AI radars and automatic aircraft fire-control radars. The consequence of this joint, coordinated effort was that the Allies would soon overtake Germany in radar technical developments—the Battle of Britain showed that they already had the lead in radar operational deployment. This lead would last until the end of the war. Today, both countries are among the leaders of radar (and sonar) innovation although, inevitably, the United States far exceeds any other country in the number of advanced radar systems that are deployed.

Sadly, this lead in deployed radar systems would develop too slowly to prevent the tragedy of Pearl Harbor, yet it might have been otherwise. Those six SCR-270 radars deployed near Pearl Harbor were state-of-the-art equipment—not the "third best" design of Watson-Watt. They formed part of the nascent AWS (Air Warning System) under the auspices of the Army Signals Corps, intended to detect and identify incoming hostile planes, scramble friendly planes, and direct them to intercept the enemy. The first part of the system (the radars) worked fine but the second part (AWS) failed. With a range of 150 miles, the radars picked up 350 incoming Japanese planes 50 minutes before the attack. The presence of numerous and suspicious radar echoes were noted and duly reported to the Command Information Center, but the officer in charge was inexperienced and untrained. He did not act on the information, and considered that the blips represented an expected flight of B-17s inbound from the U.S. mainland. This was at 7:20 a.m., some 35 minutes before the attack began. With this amount of warning, Park could have had the whole of 11-Group up at altitude, ready to dive on the approaching hostiles. The difference between AWS and the CH chain is clear. Dowding and Fighter Command in England were at war, they really believed that their system

---

[25]These useful devices were produced in bulk after the war and are still being made today. In fact, you probably own one yourself, if you have a microwave oven, since a version of the cavity magnetron provides the power for these ubiquitous modern-day appliances.

Figure 1.8 *Naval photograph documenting the Japanese attack on Pearl Harbor, Hawaii, which initiated U.S. participation in World War II. The Navy's caption: "The terrific explosion of the destroyer USS Shaw when her magazine exploded after being bombed by Japanese aircraft in the sneak attack on Pearl Harbor on Dec. 7, 1941." National Archives (photo no. ARC 520590).*

worked, and they had put a lot of effort into its development. The United States was not at war and the development of the communications infrastructure lagged woefully behind the technical development of the radar equipment; the authorities at Pearl Harbor did not take seriously what their radars were saying to them. There are lessons here that, in some quarters, have yet to be learned to this day. Some 2,400 Americans (mostly U.S. Navy personnel) were killed, and the U.S. Pacific fleet was seriously weakened. Had AWS worked properly, this loss of life and materiel could have been reduced. Also, and this point is not so widely reported, the SCR-270 radars tracked the Japanese planes *after* the attack, when they were returning to their carriers. Had AWS worked properly, then, U.S. planes could have followed, pressed an attack on these carriers, and perhaps turned the tables.

The Battle of Britain and Pearl Harbor, in the early years of World War II, serve to show how vital it was for the armed forces of belligerent nations to possess effective radar systems.

The Radar War

The CH system evolved further as the war unfolded. In the Filter Room, manual tracking of aircraft became automated. Mechanical calculators were added to perform target position calculations and these made the later CH system very efficient and reliable. Perhaps the most significant use of CH after the Battle of Britain came toward the end of the war when London was bombed from afar. Hitler sent hundreds of V1 and V2 rockets, from bases on the north coast of mainland Europe, across the English channel. Their navigation was imprecise, but London is a large target, and the warheads of these "vengeance weapons" fell indiscriminately out of the sky, causing widespread, serious damage. The CH system was adapted under the code-name "Big Ben" to detect incoming rockets—the world's first antiballistic missile system. CH tracking enabled early warning in the likely impact zone, and so reduced casualties. Backtracking provided information about the likely rocket launch site location, and then Mosquito fighterbombers would launch precision attacks on these sites and knock them out.

In 1942 the British developed the H2S radar system. These centimetric radars were installed in RAF bombers with a ground-mapping display that greatly aided navigation over enemy territory. Inland waterways, it was found, returned very weak echoes into the radar receiver, and so stood out on a radar display as dark lines, contrasting markedly with the brighter ground echoes.[26] The Germans were aware that waterways were providing navigation aids to Allied planes entering German airspace. In a desperate attempt to thwart H2S they attempted to cover large sections of rivers and canals with floating material that strongly reflected radar transmissions, so the dark lines would be filled in. This stratagem proved to be unworkable.

Centimetric radars also proved to be important during the Atlantic battle. German U-boats were very successful, for more than two years, in sinking Allied convoys supplying Britain from North America. Hundreds of thousands of tons of shipping were lost, until mid 1943 when the tide of battle turned. In part, this was due to Ultra, to improvements in the convoy sys-

---

[26]This variation in the reflectivity of topographical features is crucial to modern imaging radars, such as those onboard satellites that map the earth. Radar mapping forms the subject of chapter 5.

tem, and to other technological developments. The significant influence of radar in the U-boat war is apparent, however, from the German response. *Schornsteinfeger* (chimney sweep) was the name they gave to a graphite/rubber coating applied to U-boat hulls, intended to reduce the reflectivity of U-boats, at radar wavelengths. This early example of stealth technology worked well in experimental tests, but not when the coating became wet—something rather unavoidable on submarines. In 1944 the Germans developed the *schnorkel* so that U-boats could recharge their batteries without surfacing.[27] Long-wavelength Allied radars could not see the small *schnorkels*, or periscopes, but airborne centimetric radars could see them. U-boat losses increased significantly in 1943, effectively ending the threat to transatlantic convoys.

German radar development proceeded at a breathtaking pace throughout the war, in part due to the same pressures perceived earlier in Britain—the rising threat of air raids. However, the higher echelons of the German High Command, and the Nazi regime, were strangely half-hearted about radar. Typically, German engineers would come up with a good design, but it would not be put into large-scale production (with a couple of significant exceptions, outlined below), or not properly integrated into an air defense system. It was not until 1944 that the Germans had an integrated system as good as Dowding's. Also, the authorities were slow to recognize the potential of centimetric wavelengths—at first they did not believe that EM radiation with such wavelengths could propagate very far through the atmosphere.

The *Freya* radar, mentioned earlier, is a good example of a state-of-the-art radar that only slowly became widely available; only seven sets existed at the beginning of World War II. In 1939 the Germans developed their own cavity magnetron (though at $\lambda = 50$ cm the wavelength was five times longer than what Randall and Boot's transmitter produced). This device greatly improved the *Würzberg* ground radar developed before the war, and over 4,000 of these radar sets were to be manufactured. The Würzberg was a tracking radar utilized for flak gun laying, and by the war's end the accuracy of the Würzberg system was very impressive. Allied bombers could be detected to a range of up to 30 km with a range estimation accuracy of 10 m and an angular accuracy of 0.2°. *Freya*, with longer range and wider coverage and thus with lower accuracy, would detect incoming planes and then pass the information over to Würzberg, which tracked them and di-

---

[27]The batteries were recharged by diesel engines, which require air.

rected flak antiaircraft fire. But Würzberg could track only one aircraft at a time, so this air defense system was rather inefficient.

The Germans were also slow to appreciate the importance of airborne radar. Then in June 1941 a British Hudson bomber made an emergency landing on the northwest coast of France, near the port city of Brest, and was captured. It had onboard a sea reconnaissance radar, which was analyzed in detail. The fact that the allies possessed airborne radar spurred Germany's R&D in this direction, and by the end of 1941 they had successfully tested the *Lichtenstein* night fighter radar ($\lambda = 9$ cm). This radar did not become widely available until the following year, in part, because the authorities insisted on an internal antenna—one that is inside the night fighter airframe. This requirement proved to be impossible to implement, and eventually the antenna was mounted externally. The Allies learned how to jam[28] the Lichtenstein radars and so the Germans were obliged to produce a version of Lichtenstein with 3.3 m wavelength. The Allies had no jammer that was effective in this region of the EM spectrum, and these Lichtensteins would cause heavy losses in Allied bombers.

At first, when the United States entered World War II, the main (but by no means the only) American contribution to the radar war was to mass produce British radars. Then U.S. radar engineers began to "tweak" these designs, producing improvements. Later in the war much of the U.S. equipment was designed as well as produced at home. One important tweak was a centimetric radar (powered by a Randall and Boot magnetron) with a PPI display. PPI (plan position indicator) displays are the radar screens that lay out radar information like a map—standard nowadays, but new back then. Earlier radar displays (such as those of the Battle of Britain CH system) were simply oscilloscope traces, effective but not nearly so user friendly. Because of the short centimetric wavelengths of the radar, the resulting PPI displays were high resolution. (The H2S radar display was PPI.) Another tweak was the tunable magnetron, robust against jamming. If the Germans tried to jam a tunable magnetron radar then it would switch frequencies until it found a portion of the microwave spectrum that was not jammed. This device was used during the Normandy landings.

---

[28]Jamming falls under the heading of *Electronic Countermeasures* (ECM). This is a large topic closely related to, but distinct from, radar. I will discuss ECM in chapter 4, and include there a summary of the considerable ECM activities practiced by both sides during World War II.

One important home-grown U.S. development was the *Loran* (long-range) navigation beacon system, designed to aid aircraft flying long-distance missions. One *Loran* version had a range of 1,600 km with 4-km accuracy. As we will see, it is possible to penetrate such long distances through the atmosphere only with long wavelengths, here $\lambda = 160$ m. Such wavelengths are soaked up by the ground, however, and so this particular equipment was used only over the sea, such as the vast area of the Pacific. A different version (presumably with a different operating wavelength) eventually covered all of central Europe, guiding Allied planes wherever they wanted to go.

We have seen that radar was a decisive factor in the Battle of Britain, and was significant also in the U-boat war. It proved to be crucial in the Mediterranean during 1941–1942 and in the Pacific war from 1942 to 1943. Radar evolved to locate an enemy in the air or on water, and to navigate over friendly and enemy territory, over long ranges and without visual aids. The pace of radar development is a testimony to the importance attached to it by all belligerents—recall that this technology had been pretty much confined to the laboratory only a decade earlier. This pace would slow down once the war ended, understandably, but would continue nevertheless. Radar would find new peacetime applications: civilian air traffic control (ground control approach systems), weather system detection (as an aid to weather forecasting), ship safety (collision avoidance and navigation buoy detection), remote sensing (terrain mapping), spacecraft maneuvers (rendezvous and docking), and law enforcement (speed guns, intruder alert). I will bring you up to date on many of these applications, once we have covered the techniques that underpin remote sensing.

BIBLIOGRAPHY

Bishop, Edward. *Their Finest Hour*. Ballantine, New York, 1968.
Brown, Anthony Cave. *Bodyguard of Lies*. HarperCollins, New York, 1975. This history of the espionage war waged against and by Nazi Germany includes interesting details of the role played by radar, though the emphasis is on Ultra and the Enigma cipher.
Brown, Louis. *A Radar History of World War Two: Military Imperatives*. Institute of Physics Press, Bristol, 1999.
Deighton, Len. *Fighter: The True Story of the Battle of Britain*. Castle Books, Edison, NJ, 2000. A very readable and detailed account of the battle.
Heiferman, Ronald. *World War II*. Octopus Books, London, 1973. An American historian of World War II includes a chapter on the Battle of Britain in this well-illustrated book.

Hepcke, Gerhard. *The Radar War*. A radar history of World War II by a German radar engineer who took part in it. The text is available online at: linepc.ath.cx/topsecret/radarwar3.pdf.

Neale, B. T. *GEC Journal of Research* **3** (1985) pp. 73–83. This historical article provides a technical review of the CH radar capabilities.

Penley, W. H. www.penleyradararchives.org.uk/documents/penley/early_radar/ 15th October 2002. An interesting history of early British radar development, by one who took part in it.

Price, Alfred. *Instruments of Darkness*. Greenhill Books, London, 2005. This is a revised edition of the original 1967 history, long regarded as a standard history of the electronic war between the Allies and the Axis powers of World War II.

Skolnik, Merrill. *Introduction to Radar Systems*. McGraw-Hill, New York, 2003. This is the third edition of a textbook that has been an industry and university standard for decades.

Taylor, A. J. P. *The Second World War*. Penguin, London, 1976. Forthright views of World War II by an eminent British historian.

Wikipedia, the Online Encyclopedia, articles on History of Radar and Chain Home. (Includes quote by Maurice Ponte.)

2............................

Given that this book falls under the heading of "Technology—Radar and Sonar," I seem to have spilled quite a lot of ink on the subject of "History—Battle of Britain." I wanted to help you understand why radar technology mattered as well as what it does, and the historical imperatives that gave rise to it. Thus, when the British were seeking ways to increase the angular accuracy of their CH radar system, I guided you toward a section about amplitude monopulse. Earlier, when the NRL was conducting its wave interference experiments, I used that as an excuse to remind you about the basic properties of waves. This historical approach won't work for the remaining chapters of the book, however. (We don't know much about the history of bat echolocation evolution, for example.) So, except for a recap of World War II electronic warfare in chapter 4, I will henceforth adopt a different approach. Beginning here, I will explain remote sensing ideas more directly, unleavened by historical asides. This chapter will probably be the most challenging for those of you who are uncomfortable with math, although, as I promised, the math will be "cuddly." My aim in this chapter is to get across the key ideas as precisely as possible without math (but with enough details and with links to technical references for those of you who plan on delving deeper into our subject). By the end of this chapter, you will have gained more insight into the radar clockwork—what goes on inside and why—so that you will be able to appreciate our remote sensing apparatus at a deeper level. One example of deeper understanding, already gained, comes from note T4 (beamforming), where you saw that narrow beams require wide arrays. Next time you see a radar antenna, you will be able to infer something about its beamshape. If the antenna is wider than it is tall, then you know that the azimuth beamwidth is narrower than the elevation beamwidth. (You can't be more precise without knowing the transmitter wavelength.) So, this radar antenna is intended to provide information about the bearing direction of whatever targets it seeks.

I will provide you with an appreciation of the challenges faced by radars; for example, how those targets have different and variable echoes, which are much, much fainter than the "background" echoes. Also, we will dip our toes in the waters of sonar remote sensing; the challenges faced by submarine sonar are very similar to those of radar, but the environment is so different that sonar technology has developed special adaptations to cope with it. One of my main goals in this chapter is to get you to the radar range equation. This is a simple but crucial formula that provides us with an estimate of the maximum detection range of a radar system. (The same equation works for sonar range, though sonar systems engineers usually express it differently.) Before going there, though, we need to learn some concepts, such as radar cross section, antenna gain, and background noise. These and other factors appear in the range equation—hence, the preliminaries.

dB or Not dB—Decibels and Other Preliminaries

You cannot get very far in engineering before encountering the ubiquitous decibel (dB). This is a unit of measurement, or rather of relative measurement—how big or small a gizmo is, when compared with a standard gizmo. Decibels are becoming a part of everyday life as technology increasingly permeates every aspect of it. We have all heard of decibels, but they are often poorly understood. Hence note T8, which provides you with all you need to know about decibels.

The road to the radar range equation begins with note T9, in which the notion of antenna gain is explained. (Now that you are experts at decibels, I will use them in note T9 to compare numbers.) Put simply, antenna gain is the increase in power along the direction of the antenna mainlobe (the main beam, recall). A highly directional tracking radar antenna (one with an array length that is much greater than the wavelength of radiation that it transmits or receives), with a pencil beam, has a higher gain than an antenna belonging to a surveillance radar. The surveillance radar needs to search a large volume of sky and so has wider beams. The tracking radar needs to pinpoint a particular target, such as an incoming missile, and so wants to exclude extraneous signals. Also—as the radar equation will show us—a high-gain antenna leads to greater radar range.

So now we have a transmitter that is sending out microwaves in a particular direction—toward a target. How much of the microwave power gets there? Quite apart from simple geometrical spreading—the range equation will take care of that—how much of the transmitted radiation is absorbed by the atmosphere, or scattered by molecules and particles within the atmosphere? This is the subject of note T10. A lot of experimental and theoretical work has been put into atmospheric attenuation of EM radiation precisely because of its significance to radar detection range. As you will see, certain wavelengths are absorbed very strongly. The atmosphere simply soaks them up, or scatters them in all directions, whereas other wavelengths cut through the air as if it wasn't there. This difference provides an important constraint on radar design, as well as a limit to detection range.

The last preliminary hurdle to be jumped is *radar cross section*, or RCS. There are some aspects of RCS that are not obvious, and are worth emphasizing in a separate note, hence, note T11. The basic idea is very simple, however: the RCS of a target is the area it presents to a radar sensor. The devil, as with much of life and all of engineering, is in the details.

The Radar Range Equation

We will calculate the power that is returned from a target back to the radar receiver, in terms of the transmitted power. We can do this by following the microwaves out of the transmitter, through the air to the target, and back again. First, let us denote the power transmitted by the radar power unit, be it a magnetron, Klystron, traveling wave tube, or whatever, by P. For an airborne radar P might be several kilowatts; for a large land- or ship-based radar it may be a hundred times bigger. For a small mm-wave tracker radar, P might be only the power of a light bulb. Now our microwaves are directed by the transmitter antenna toward the target. We have seen that this concentration of power in a chosen direction is expressed by a gain factor, here written G. Without gain, the power would spread from the transmitter uniformly in all directions, and by the time it had traveled a distance R from the transmitter, the power density (power per unit area) would have dropped to $P/4\pi R^2$. The denominator here is simply the area of the surface of a sphere of radius R. Including gain, the power density is

$$P\frac{G}{4\pi R^2} \quad \text{Power density at range } R$$

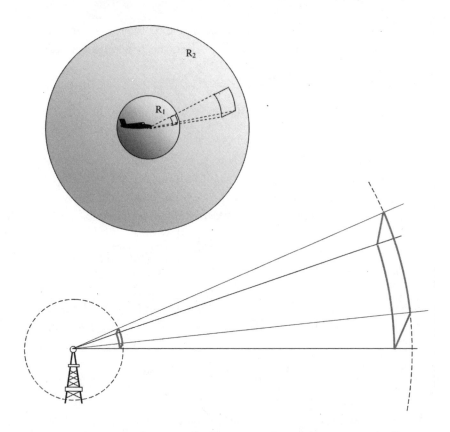

Figure 2.1 *Two views of square radar beams (unrealistic, but convenient for illustration). They intersect two imaginary spheres with radii $R_{1,2}$. The intersection area increases as the square of the radius, so the radar transmitter power density falls as the inverse square of radius.*

Imagine the microwaves spreading out from the transmitter, as suggested in figure 2.1. The power density is "diluted" as R increases, but is increased by a factor G in the target direction, due to antenna beamforming. The overall effect is to produce a power density at range R of PG divided by the area of the spherical surface, as shown. If this is not clear to you, then look at figures 2.1 and T9.1 for a while, and you will see that it makes sense.

So if a target is at range R, we know the power density that illuminates it from our radar transmitter. The actual microwave power striking the target is just the power density multiplied by target area—the RCS. Now RCS is usually denoted by the Greek letter σ, and so our radar has hit the target with power

$$P\frac{G}{4\pi R^2}\sigma \quad \text{Power striking the target at range } R$$

This power then reflects off the target in all directions—imagine another expanding sphere. Back at the radar receiver, the power density reflected from the target is

$$P\frac{G}{4\pi R^2}\sigma\frac{1}{4\pi R^2} \quad \text{Power density reflected back to the radar}$$

Note that there is no gain factor here, since the target is assumed to scatter in all directions equally. (We know that this isn't true, but on average . . .) The amount of reflected power that is picked up by the radar receiver is just the power density multiplied by the receiver antenna area A, so

$$P\frac{G}{4\pi R^2}\sigma\frac{1}{4\pi R^2}A \quad \text{Power received by the radar}$$

Now putting this all together and including a loss factor L for attenuation through the atmosphere,[1] we have one form of the radar range equation:

$$S = \frac{PG\sigma AL}{16\pi^2 R^4} \quad \text{Target signal power} \qquad (1)$$

So S is the power reflected from a target and picked up by the receiver. You will appreciate that S is an average power, given the nature of the calculation. We know that target RCS varies with look direction (and will soon see that it also fluctuates in time) and, to obtain a general range equation, we use the average RCS value. Also we have seen that losses are not precisely calculated, due to localized atmospheric conditions, and so we assume "average" atmospheric conditions when determining L.

Now let's see what the range equation is telling us. Signal power increases if we increase transmitter power or target RCS—unsurprisingly. It falls rapidly as the fourth power of range, however, and this is the killer. Double the range, and target signal power drops by a factor of 16. The amount of power returned from a distant target is *miniscule*. To demonstrate how tiny the signal power is, let us put some numbers into equation (1) and work out the ratio of signal power S to transmitted power P. If the gain is 30 dB and the radar is a 5-m^2 fighter plane (RCS +7 dB) detected at range

---

[1] In most textbooks this loss factor is expressed as a gain factor (i.e., L exceeds one) in the range equation denominator. I have never understood why this convention persists.

$R = 100$ km, and if our receiver antenna has an area of 1 m² (or 0 dB) and the loss factor is -5 dB (all plausible numbers for real radar systems) then we find that $S/P = 0.000000000000000001$ or -190 dBc. How in the name of all that is wonderful can we pick up such a small echo signal? If the transmitter power is $P = 1$ kW (30 dBW) then the target signal power is one ten-thousandth of a micro-microwatt. This isn't enough to power a bacterium,[2] let alone show up in a radar signal processor. It is certainly much less than the noise power inside a receiver.

Cue another technical note on electronic noise? Well, I thought about it, but on balance I reckon that you will have been through enough by the time you emerge triumphant at the end of this chapter. So, in a nutshell, electronic noise can be regarded simply as unwanted power, at all frequencies, due to the heat generated by electronic components. Some noise is not due to heat, but can be treated as if it is—the receiver is assigned an "effective temperature" from which its noise power can readily be found. The chief characteristic of noise power is that it fluctuates randomly from one instant to the next (in a manner to be discussed shortly), so that a sample of noise power is completely independent of the previous sample. Noise also is spread evenly over the receiver band. So, a wider bandwidth means more noise power.

The teeny signal power gets swamped by the internal noise power of the radar receiver and signal processor from the instant it enters the receiver antenna. For radar systems engineers, the key factor is *signal-to-noise ratio* (SNR), rather than the absolute signal power[3] S. If the noise power of a particular radar system is N watts then, typically, S/N might be -30 dB. Here S is the power of the target as calculated from (1). So why isn't the signal simply lost amid all this noise? To illustrate the problem, I show simulated noise and a small signal in figure 2.2.

Most of signal processing—the clever algorithms and techniques that form the subject matter of chapter 3—is aimed at making the signal visible. Signal processing boosts the tiny signal power S so that it stands out above the noise N and is detected. You can begin to appreciate the magnitude of the task from figure 2.2, when you realize that the signal there

---

[2] I'm guessing—I haven't asked one.

[3] This is because a small signal—even the -190 dBc target echo just calculated—may be detected quite easily if no other signal is present. A minute blip on an absolutely flat background can be detected by simply amplifying the signal. Of course amplification does not help if there is a competing noise signal, because both target echo and noise will be amplified.

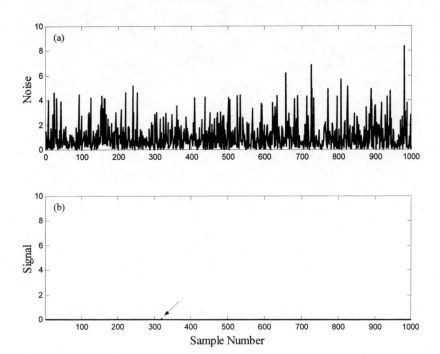

Figure 2.2 (a) Simulated electronic noise, fluctuating about a mean power of 1. (b) A target signal in the 320th sample, with power 0.1, that is, 10 dB below noise.

---

has one-tenth of the noise power (SNR = −10 dB), which is 100 times larger than the "typical" value of −30 dB that I quoted above. Also, noise is only part of the undesirable interference power that hides our target signal. A large amount of transmitter power is also echoed back to the receiver from sources other than our target—from the environment. This is radar *clutter* (sonar *reverberation*), and clutter is the subject that I now need to convey to you.

Clutter in the Workplace

I will guess that more radar-engineer hours of research and experimentation have been invested in understanding the many and varied forms of clutter than any other single radar topic. Clutter is a many-headed monster that crops up everywhere you look. Indeed, clutter is almost all of what a radar looks at, or listens to. Clutter, in short, is everything that the radar detects except for the target. Now, imagine an army surveillance radar that is searching a volume of sky for an incoming missile. The missile may occupy 2 m³ of sky, whereas the search volume may be $10^{13} = 10,000,000,000,000$ m³.

The radar echo from within the 2 m³ is signal; the echo from everywhere else is clutter.

Before describing clutter I will emphasize an important difference between clutter and noise. Noise arises inside the radar, and so is completely independent of what is outside. Clutter arises from radar echoes from the environment, and so depends on that environment. In particular, clutter varies with range, and look direction, whereas noise does not. (Noise doesn't vary with anything, except temperature and time.) The sum total of noise power plus clutter power is called *interference*, and it is the interference that our target signal is competing against, inside the radar or sonar signal processor.

For the purposes of classification, clutter can usefully be broken down into three main groups, though each group may contain clutter sources of many different stripes. In this chapter I will label the clutter groups numerically, and the number applied will be significant, as you will see. First, our army surveillance radar (and the CH system of chapter 1) had to deal with *volume clutter*, here denoted as type 1. Examples of volume clutter (reverberation) include clouds, rain, falling snow, insect swarms, bird flocks, chaff, trees . . . (bubbles, shrimps, shoals of fish, kelp forests . . .). Radar engineers have measured the RCS of insect swarms, and of birds. I once read a technical article that reported measurements of the radar cross section of different types of birds, viewed from different angles.[4] Rain and clouds, in particular, have been investigated with great thoroughness, because, in the past, these meteorological phenomena have seriously limited the airborne operations of the military.[5] What matters with volume clutter is not so much the RCS of an individual starling or shrimp or raindrop, but the cumulative effect of large volumes of these starlings, shrimp, or raindrops. For this reason, type 1 or volume clutter RCS is more meaningfully specified as RCS *density* rather than as straightforward RCS. RCS density is RCS per unit volume. Thus rain clutter, for example, is specified in the rather odd units of square meters (i.e., RCS area) per cubic meter: m²/m³. When a radar engineer needs to calculate the clutter RCS due to rain, he

---

[4]Ducks and geese have a relatively large RCS when viewed at centrimetric wavelengths from the side or front, you will be interested to learn, because of their prominent beaks.

[5]One reason why, in World War II, the Germans chose to attack U.S. forces at Bastogne during the winter (1944–1945) was that cloud cover prevented Allied planes from attacking at that time of year. (During the last few months of the war, the Allies had complete control of the skies, and so the Luftwaffe could not provide an air screen for land operations.) Later, in the 1990s, coalition forces found that operations over the former Yugoslavia were more difficult than operations over Iraq, in part, because of clouds and rain obscuration.

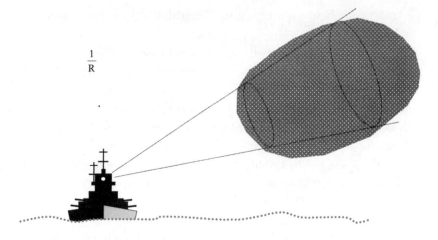

$\dfrac{1}{R}$

Figure 2.3 *Type 1 clutter: volume clutter—here a rain cloud centered at range R.*

takes the measured RCS density for raindrops and multiplies by the volume of rain seen by the radar.

Needless to say, rain RCS density depends on rainfall rate, but it also depends on radar wavelength. Thus, if you consult one of the myriad of clutter graphs in radar textbooks, you will find that for $\lambda = 3.2$ cm the RCS density for rain falling at the rate of 10 mm/hour (a moderate shower) is about 10^{-6} m²/m³. This may not look like much, but it becomes significant if the radar is covering a large volume of rain. The main reason why so much engineering effort has gone into clutter measurement is because of the varied form of clutter, and because the RCS values obtained vary with frequency and other factors.

I refer to volume clutter as type 1 because of the way in which it changes with range. Consider figure 2.3. Here is a warship that is concerned about enemy airborne threats, such as dive-bombers or missiles. In searching for or tracking an enemy target, the ship radar must penetrate a rain cloud, as shown. Now we know from equation (1) that target power depends on range as R^{-4}, but this is not the case for volume clutter. Sure, we might say that the center of the rain cloud is at range R but, as R increases or decreases, the volume of rain clutter *within the radar beam* also increases or decreases. So the total rain clutter RCS varies like R^{-4} (range equation) multiplied by R^3 (volume), that is, like R^{-1}. So volume clutter falls away slowly with range. I label it type 1 clutter because it decreases with range as R^{-1}.

Type 2 or *surface clutter* varies as R^{-2}, whereas type 3 clutter behaves like R^{-3}. I will refer you to note T12 for enlightenment on these types, and on the complications that arise for signal processing when clutter is present. Note that, because clutter falls with range whereas noise does not, we can expect that clutter dominates the radar echoes from short ranges, whereas noise dominates at longer ranges. The crossover range, when noise and clutter contribute equally to the interference power, separates *clutter-limited* and *noise-limited* regions, in the jargon of radar systems engineers. Note also that target signal power falls faster with range than does clutter, whatever the type, so that longer ranges always mean lower signal-to-interference ratios, or more difficult target detection.

Let's Hear It for Sonar

This book is about sonar and echolocation, as well as radar, because both acoustic and EM transmitters are utilized for remote sensing. So far I have concentrated overwhelmingly on radar because my historical approach emphasized the sudden flowering of remote sensing brought about by radar development before and during World War II. Much of the signal processing utilized by sonar systems is the same as that of radar systems, but the radically different environments in which these systems operate has meant that, to a considerable extent, the development of radar and sonar has progressed in parallel, rather than arm in arm.

The first question that needs to be addressed is this: why bother with acoustic radiation at all, when EM radiation works so well? The answer is provided in figure 2.4. EM radiation is absorbed by seawater much more strongly than it is by air. The fraction of EM power absorbed as it travels through 1 meter of water can exceed 99.9999 percent, depending on frequency. As you can see, there is a narrow window in the visible portion of the EM spectrum; here as little as 2 percent of the power is absorbed per meter, so that visibility extends to several tens of meters at these wavelengths. It is no coincidence that the narrow band of relatively penetrating EM radiation is precisely the band to which our eyes are sensitive. We evolved from sea creatures, and our vision evolved in the sea. It would not have evolved at all if there were no such window, since it would confer no evolutionary advantage in that case.

Note from figure 2.4 that at very long wavelengths the penetration of EM radiation increases (absorption is reduced). This fact permits communication by radio waves with submerged submarines. However, the long wavelengths (low frequencies) mean that the communications signal is of narrow bandwidth, and so data transfer is at a very slow rate. Also, such very long wavelengths require long transmission antennas, and if these are

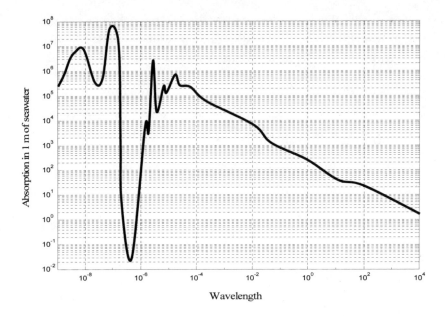

Figure 2.4 *Absorption of EM radiation per meter of seawater. Absorption of 10^6 means that only one part in 1,000,000 of the incident radiation penetrates 1 meter. Note the relatively low absorption at visible wavelengths and at long radio wavelengths.*

submerged in a conducting medium such as saltwater, then they hardly function at all. For this reason, long-wavelength radio communication with submarines is a one-way process: headquarters on land can communicate with their submerged submarines (at a slow data rate) but the subs cannot reply.

Acoustic radiation, on the other hand, penetrates water comparatively well, in fact, better than it penetrates through air. This is because of the relative density of water compared with air. The speed of sound through water is about 1,500 m s^{-1}, or about five times faster than in air. A powerful acoustic transmitter (called *transducers*—though I shall, for simplicity, refer generally to all remote sensing wave sources as transmitters) can send and receive signals many kilometers. In certain special circumstances, as we will see, communication can extend hundreds or thousands of kilometers, practically girdling the globe. Sound is absorbed by seawater, but only at very high frequencies—above 1 MHz. For frequencies below 1 kHz the absorption of sound waves is less than 1 percent per *kilometer*. So, in summary, sonar will work underwater, whereas radar will not, because of absorption by seawater.

There are difficulties in underwater sonar operation that are absent in working with airborne radar. The sound waves do not travel in straight lines. Because of temperature and salinity gradients, in particular, near the ocean surface, acoustic rays bend. To be sure, light rays and microwave rays bend in air, (for example, due to varying air density with altitude), but this is usually a small effect. I shall defer a discussion of these anomalous propagation effects until later, but for now I ask you to note that the effects are far more severe for underwater sonar than for radar.

Another significant difference is that there is a lot of *external noise* in the ocean. Earlier I defined noise as being unwanted random signals originating *within* a radar system. There is equivalent noise within a sonar system but, in addition, there is noise from the ocean itself. This is not the same as reverberation (unwanted echoes from the environment). Life within the ocean generates its own noise, and this can be a significant factor in sonar operation. I mentioned copulating shrimps earlier—a serious problem at certain times of year for submarine sonar operations.[6] Many other creatures make many other noises at all times of year. These noises, and nonbiological noise such as breaking surf, rainfall on the ocean surface, and sea ice "growling" contribute to external background noise.

Submarines themselves generate a variety of noises. There is something called *flow noise*, which originates from turbulence in the vicinity of *hydrophones* (underwater microphones—again for simplicity I shall henceforth use the radar term: receivers). There are characteristic pressure waves (very-low-frequency sound waves) that result from large objects, such as submarines, moving through the water. *Cavitation* is a phenomenon that plagues submarines. A rotating propeller causes bubbles to arise in the wake of each blade. At depth these bubbles are extinguished quickly by water pressure, and are snuffed out with a characteristic snapping sound. Cavitation is bad for two reasons. First, the sound gives away the presence of a submarine, should a sharp-eared enemy be lurking nearby. Second, the cavitation bubbles snapping shut is extremely violent, though highly localized, and over time causes pits and other damage to the (very, very

---

[6]Not to mention farting herrings. The Swedish Navy, always with a sharp ear for Russian submarines, was once obliged to investigate suspicious rasping sounds that were being picked up from deployed sonar systems. Much to the navy's surprise they found herrings; well, so reported the *Guardian* in England. It seems that herrings fart more when in shoals than when alone, as a form of communication. Doesn't it just make your day, to be privy to such arcane knowledge?

expensive) propeller. Many engineers have spent entire careers designing quiet submarine propellers.[7]

A third difference—this time a blessing, in general, rather than a curse—is that acoustic radiation can travel through a target, as well as reflect off it. This property of sound waves means that sonar signal processors have more information to work with when trying to identify a target. A radar signal processor receives echoes from the outer skin of its targets, whereas the sonar processor receives echoes from the outer skin and from internal structures. Additionally, the target will often emit its own characteristic sounds. Such passive radiation is extremely important and useful, as we will see.

There are other differences between sonar and radar operation, in particular, the disparity of physical equipment, but I do not want to emphasize these differences, because this might give the impression that sonar and radar are distant cousins. On the contrary, I wish to show that they are siblings; the differences are interesting and important, but sonar and radar both work in the same family business of remote sensing.

For now I will merely touch base with echolocation, which is the non-human application of sonar. Echolocation has evolved in many different species, but in only a few groups, of animals. The most sophisticated echolocators are mammals, and I will concentrate on these in this book. Underwater echolocation is utilized principally by cetaceans—certain whales and dolphins. Airborne sonar is the realm of bats, and in particular, of the small microchiropteran family. Both groups have adapted beautifully and in minute detail to the use of echolocation in their particular environments. They deserve a section of their own here, but that comes later. For those readers who bought this book principally to learn about echolocation, please compose yourselves in patience. You do not yet have all the prerequisites to keep pace with these creatures—after all, they have had several tens of millions of years in which to learn about remote sensing and signal processing. We human neophytes are catching up fast, but you are still separated from bats and dolphins by at least a couple of chapters.

---

[7]During the Cold War, western propellers (on U.S., British, German, and Dutch) hunter-killer submarines were quieter than their Soviet counterparts, and so western subs were less easily detected. This, of course, was a significant tactical advantage, which ended in the mid-1980s, when the Walker spy ring in the United States sold key design features to the U.S.S.R. Thereafter, Soviets subs were also very quiet. Happily there were no military disasters as a consequence of this nefariousness, because the U.S.S.R. itself did not survive the 1980s.

A Small Sampling of Statistics

If you don't like math then you will hate statistics. But please persevere without sulking: the study of statistics is good for you even though it is difficult to swallow, like cod liver oil. In note T13 you are introduced to the basic vocabulary and concepts needed for this book. A few examples will be given to aid comprehension, and a few misconceptions corrected and bubbles burst (without the violence of cavitation).

Why bother? Because remote sensing is an inherently statistical subject. This is because the environment contains a lot of random processes, which scramble the orderly transmitted pulses. It is as if a cohort of well-drilled cavalry, newly kitted out in spanking clean uniforms, trots out of the fort in formation, swords gleaming, to take on a foe, and then returns piecemeal and cut up, bedraggled and incoherent—barely recognizable. This mutilation of our neat and tidy transmitted pulse train comes about because the environment that reflects the pulses back into our receiver is not neat and tidy. Consider—this is just one example of thousands—what happens when the pulses echo off a tree. The leaves of the tree face all directions and some of them will give rise to much larger echoes than others. If the leaf is facing our radar transmitter, it will possess a larger RCS than if it were edge on. Also, different leaves are at different ranges from the radar, and so the echoes will return at different times. Now if a breeze is blowing the leaves will move in a manner that for all intents and purposes is random, so the next pulses will see a different tree. So, the echoes from clutter objects can be—usually are—random and fluctuating.

The same is true of most targets. Some simple targets do not fluctuate and have a fairly simple and consistent echo. However, a complex and multifaceted target such as the fighter plane of figure T11.1 has an RCS that changes significantly with look direction. So if such a target turns, even a little bit, during the time it is illuminated by a radar beam then the echo will change. Another factor that tends to randomize the echoes, of both clutter and targets, is atmospheric turbulence and variability. We have seen how the attenuation of radar signals traveling through the atmosphere depends on range—but the absorption and scattering effects we discussed earlier account only for the average attenuation level. In practice the attenuation for a train of radar pulses fluctuates about this mean value, from one pulse to the next, due to localized and rapidly changing atmospheric density along the path traveled by our pulse train.

The distribution of echo power depends on what type of target, and what type of clutter, the echo comes from. In the 1950s a lot of theoretical work,

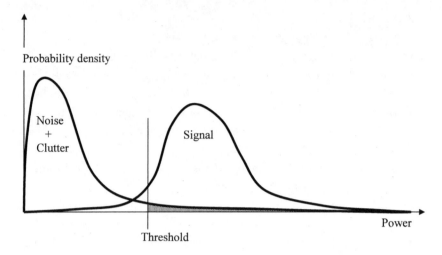

Figure 2.5 *The statistical distribution of interference power (clutter plus noise) and of a strong target signal. If an echo pulse exceeds a specified power threshold, then a target is declared. This inevitably leads to false alarms, because some clutter samples may exceed threshold (shaded area). A fundamental trade-off exists between target detection probability and false alarm probability.*

principally by Marcum and Swerling in the United States, showed that targets fall into one of five distinct types: one that has a constant, unvarying RCS and four with fluctuating RCS. The distributions of the fluctuating targets are different for each type. For clutter fluctuations the distributions are, not surprisingly, given the enormous variety of clutter objects, much more varied. Radar engineers, since the earliest days, have exerted a huge amount of effort in observing, classifying, and understanding the nature of clutter fluctuations. Some of the distributions are very noiselike and common. Others are very complex—sea clutter fluctuations fall into this category—and are still being actively investigated.[8]

Given that noise fluctuates and that clutter fluctuates, you would expect that radar interference (the sum of noise and all the various clutter sources) also fluctuates. Indeed so. In figure 2.5 I sketched a typical interference

---

[8] I once reviewed a book for the *Journal of the Royal Aeronautical Society*. It was the latest edition of a long-running and successful book about radar clutter, written by an American radar engineer (Long) who had made his name in this field. There were literally hundreds of graphs and tables summarizing different clutter characteristics and fluctuation properties. So far as sea clutter is concerned, you may (or may not) be interested to learn that it fluctuates in a manner that very closely mirrors the way a star twinkles at night.

power distribution. There is a single sharp peak and a very long tail at the high-power end. Now imagine a radar echo from the same clutter, plus a target. This will have a different distribution, due to the presence of the target. For a strong target that echoes more power than all the clutter and noise combined, this distribution of "target plus interference" will have a distinct peak, as shown. If the signal power is small compared with the interference power (and we saw in figure 2.2 that it usually is, prior to signal processing) then the difference between target plus interference and interference distributions will be negligible.

Let us assume that, through techniques discussed in chapter 3, we have succeeded in boosting the target power relative to interference power, so that the distribution peaks are distinct, as in figure 2.5. Now it is a simple matter to tell when our echo contains a target signal. If the echo power exceeds a certain threshold value—which we can calculate if we know the target and interference distributions—then we simply say that the echo contains a target if the power exceeds our threshold, and it does not contain a target otherwise. In other words, we are using our knowledge of noise, clutter, and target statistics to set a threshold power, and declare a target to be detected if an echo exceeds this threshold. We "test" our echo by applying a threshold. This is the usual way that targets are discriminated in radar and sonar systems.

But there is a problem that is unavoidable and plainly visible in figure 2.5. Wherever we set the threshold, the possibility exists that an interference echo will exceed it. This means that we will declare a target to be present when in fact there is no target—just a particularly strong interference sample. This phenomenon is known as a *false alarm* and it must be very carefully accounted for, when processing the echo signals.

Where should we set the threshold? If we increase the threshold power to reduce the probability of false alarm, then we also reduce the probability of detecting the signal. In some situations it may be acceptable to miss a signal—for example, if we expect many echoes from the target, any one of which might lead to its detection. In other situations we cannot afford to miss a target—for example, if it corresponds to a fast-approaching missile. Suppose we lower the threshold to increase our chances of detecting the target. In this case we also increase the chances of false alarm. Again, in some situations this may be acceptable. For example, if we were tracking a target for many minutes, then the occasional false alarm on our radar screen would not matter, since it would probably be in a different location from our target track. In other situations, false alarms are completely unacceptable. For example, if our remote sensor monitored the inside of a fighter plane cockpit, and launched the pilot ejector seat when it thought a smoke threshold or temperature threshold had been exceeded.

In radar detection theory the procedure adapted is what statisticians call the *Neyman-Pearson criterion*. This provides an optimum method for deciding whether a signal is present in the radar or sonar echo. The idea is to acknowledge that we will always have false alarms, wherever we set the threshold, then decide how many false alarms are acceptable, and then maximize (by appropriate signal processing) the probability of detecting the target when we test our echoes. So, if we judge that our tactical situation permits a false alarm rate of 10^{-5} (corresponding to one false alarm for every 100,000 tests) then we set the threshold at a level to produce this number of false alarms. For a large SNR or SIR (signal-to-interference) power ratio, as in figure 2.5, the target detection probability for this threshold level may exceed 90 percent. For smaller SIRs the detection probability decreases. This trade-off between false alarms and detections is a fundamental one, in all radar and sonar signal processing.

I will close this chapter with a brief summary of what we have accomplished, and what yet needs to be done. After introducing some fundamental radar concepts such as RCS and atmospheric attenuation, we derived the radar range equation. From this equation it became clear that the signal power returned to a radar receiver is very weak. The weak target signal is lost in noise and clutter, and we looked at the many and various sources of clutter that constitute much of a radar echo signal. An introduction to signal processing statistics led us to detection theory: *if* we can boost the target signal so that it is stronger than the interference signal, then we can set a threshold and test our echo pulses to see if they include a target. The signal processing methods that we have learned over the years, of boosting the target echo, are presented in chapter 3.

Finally, everything I have said in this summary about radar target detection applies to sonars. We have seen that there are some peculiarities unique to underwater sonar, arising from the very different environment, but we can anticipate that the signal processing techniques to which I now turn will apply in both cases. The signal processing strategies of nonhuman echolocators can only be properly appreciated after we have seen what humans have learned about this subject.

BIBLIOGRAPHY

There is a paucity of accessible literature on remote sensing (hence, this book). The same cannot be said of engineering books and technical papers: the difficulty here is in choos-

ing the most readable. Almost all my technical explanations and examples for this chapter and the next come from notes accumulated over 20 years or so. Here, for the diligent student, are textbooks and course notes that I (and many others in the field) have found to be very useful. And the farting herring article.

Atkins, P., and Creasey, D. J. *Underwater Acoustics and Sonar Systems.* Department of Electronic and Electrical Engineering, University of Birmingham, England, course notes, April 1987.

Barton, David. *Radar System Analysis.* Artech House, Boston, 1976.

Guardian. 2005. Farting fish fingered. March 11.

Long, M. W. *Radar Reflectivity of Land and Sea,* 3rd ed. Artech House, Boston, 2001.

Skolnik, Merrill. *Introduction to Radar Systems.* McGraw-Hill, New York, 2003.

3 •••••••••••••••••••••• 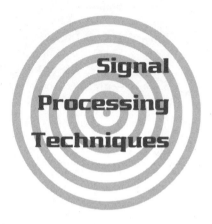 Signal Processing Techniques

Our first order of business is to examine several signal processing schemes that greatly enhance the target signal, relative to the unwanted noise and clutter signals. Once this has been achieved, then we know that a carefully chosen threshold can be applied to tell us whether a target is present. When we have got to that stage—target detection—then we can move forward and look at sophisticated methods for learning more about the target. The most important information, in many radar and sonar applications, is a precise knowledge of target position, heading, and speed. We will see how to achieve these goals so that, by the end of the chapter, we will have a remote sensing system capable of detecting and tracking many targets, even when they are moving in a highly cluttered environment.

So far we have seen how radar and sonar remote sensors can detect targets and find their approximate positions. The accuracy of these position estimates is limited by the accuracy of estimating target range, azimuth, and elevation. We say that the target position is specified by the sensor to within a *resolution cell* (or *box* or *bin*). Thus, metaphorically, the radar or sonar has placed the targets inside a box—this is illustrated in figure 3.1. At long ranges, or for low-accuracy systems, these boxes may be quite large—several kilometers across. We will see later how position estimation accuracy may be improved so that the target location can be pinned down to a box that is far smaller than the basic resolution cell. (In chapter 1 we saw how

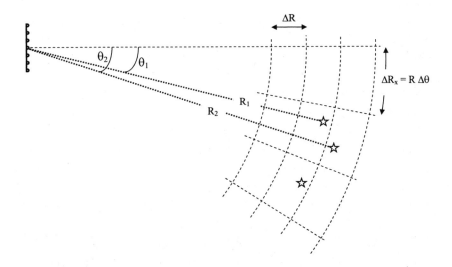

Figure 3.1 *An array antenna divides the search volume into resolution cells. The cell dimensions depend on the radar or sonar range estimation accuracy ΔR and angle estimation accuracy (here, azimuth accuracy $\Delta\theta$). Note that the width of the box increases with range. Targets (stars) are resolved if, as here, they are in separate boxes. So, the top two targets are seen separately because they are in different range cells, even though their angular separation is small: $\theta_2 - \theta_1 < \Delta\theta$. Similarly, two targets in the same range cell are separated by angle.*

amplitude monopulse was used to improve angular accuracy beyond the azimuth estimation capability of the CH beam. Monopulse techniques are still used today for this purpose.) Even without advanced techniques, placing targets within resolution cells is of considerable value for clutter rejection. For example, if we want to boost the signal-to-interference (SIR) power ratio of a particular target, and the range equation has taught us that this will usually be necessary, we can reject all the clutter that originates from outside the target resolution cell. This means that clutter from the entire volume of a radar beam—perhaps 1,000 km^3—does not compete with the target signal for our attention. Instead, only the clutter within the target box competes—this may be 10 km^3 or it may be 0.001 km^3, depending on radar position estimation accuracy, but whatever it is, it is much smaller than the volume of space illuminated by the radar beam.

We obtain significant further improvement of SIR by employing *Doppler processing*. The phrase "Doppler radar" appears now to be in general usage—it is casually dropped into many weather forecasts, for example, especially those showing images of large hurricanes and other meteorological systems—but without any sensible explanation that I have seen. Doppler processing is, in fact, central to all modern remote sensing systems, whatever their application.

Let's Be Coherent: How the Doppler Effect Helps Remote Sensing

Modern radars have progressed beyond the modest aims—compelling enough at the time—of providing basic information about the existence and general location of invading Luftwaffe formations or of Japanese dive-bombers. Nowadays we are much more ambitious. We want to know—and new technology has enabled us to find—the position of an enemy with vastly improved accuracy. Our enemy will not want us to find him[1] and will do whatever he can to reduce his RCS. Modern radars, though, can see much smaller targets than their World War II ancestors could see. They can see smaller targets further away and pinpoint them more accurately, but there's more. Modern radar can describe an airborne target—fixed wing or rotary, number of engines, airspeed, heading. They are beginning to be able to *classify* the target type (fighter, bomber, etc.) and will soon be able to *identify* targets (F16 fighter, backfire bomber, hang glider, and so on). These advances in radar techniques will be the subject matter of chapter 6. Radars cannot yet tell us what the pilot had for breakfast, but given another 50 years . . . The big difference between radar systems today and those early devices of chapter 1—a difference that permits such advances to be made—is that modern radars are *coherent*. Coherence is explained in note T14. Coherence, in this technical sense, is essential not only to military radars but also to the most important civilian application of radars, that of image formation or mapping. In essence, a radar waveform is said to be coherent when successive pulses are carved out of a single wave rather than from different waves—like a wall formed from a solid block rather than from a multitude of bricks. Coherence allows us to do Doppler processing, to which we now turn.[2]

---

[1]A radar reflector that does not want to be observed, such as an enemy bomber, is a *non-cooperative target* in the jargon of radar engineers. This contrasts with a civilian airliner, for example, who most definitely does want to be seen on air traffic control radar screens. Such *co-operative targets* will be the subject of a later chapter.

[2]Again, purists will raise their hackles at this statement. In fact, crude Doppler processing of noncoherent radar signals is possible. However, the impressive performance of modern *pulse-Doppler* radars arises from their coherent signals, which permit frequency-domain signal processing and lead to exquisitely fine angular resolutions and sharp radar images.

Perhaps you are a chickadee devotee, a plover lover, a raven maven—in short, like me, a bird nerd? Or, almost as likely, a duck hunter? (Stay with me—I have not blown a neuron—this is relevant.) The burgeoning number of people who fall into one or both of these categories (and they are not mutually exclusive) have an appreciation of bird flight and, in particular, an admiration for their migration skills. How can a barn swallow find its way south in the winter, across the equator, and then find its way back to the very same nest site next spring? We are learning a lot about bird remote sensing and navigation. It does not quite fall into our radar and sonar/echolocation family; it is more like first cousin. Birds make use of environmental clues—rivers, coastlines—and this is something that we humans are picking up on and incorporating into our remote sensing arsenal, as you shall see in chapter 6. Birds have adapted exquisitely to two things: flight and navigation. For the latter they have learned to soak up all the information that comes their way from the environment. This is passive sensing, as defined in the Introduction. Birds use the visible part of the EM spectrum about as well as we do, but they have squeezed more out of other sources of environmental information. Some birds can hear lower frequencies than we can, and this is thought to help them navigate at night or in bad weather when vision is impaired—they can follow a coastline by hearing the low-frequency waves coming ashore. Other birds (like vultures) follow scent gradients to home in on food. Many navigate by the stars, and several species (e.g., pigeons) have a sixth sense unavailable to us: they can "see" a component of the earth's magnetic field, and apply this to help them on long north–south migration routes.

I propose no further discussion in this volume about these fascinating creatures. My purpose in mentioning birds' navigation skills is to emphasize the need to make use of *all* the information that is available, in whatever form, and exploit it. Early on in the development of human remote sensing, R&D engineers and scientists learned to exploit a well-known phenomenon of physics. Perhaps more than any other single phenomenon the Doppler effect has improved our ability to pick out faint signals—the essence of remote sensing.

Doppler processing literally adds an extra dimension to the way in which we detect and enhance target signals. Doppler allows us to estimate target *speed*, and so we can discriminate targets (separate them from the surrounding clutter) in four dimensions: range, azimuth or bearing angle, altitude or elevation angle, and now speed. Strictly speaking, Doppler processing provides us only with information about the radial component of a target velocity, that is, the component toward or away from the radar or sonar sensor. Any tangential component is not measured. This is a consequence of the Doppler phenomenon, a physical effect that is summarized in note T15. Basically, a moving object changes the frequency of a signal, so that the echo signal is shifted slightly[3] in frequency compared with the transmitted signal. Doppler processing estimates this frequency shift and from the frequency estimate we determine target speed.

It is pretty obvious that knowledge of target speed is a good thing; for example, in a military context you would like to know how long you have got before a missile gets to you. Weather radars rely on Doppler processing to provide accurate pictures of wind speed and hurricane formation, etc., as we will see later. But Doppler information also enables us to enhance SIR by reducing the amount of clutter, and even of noise, that competes with a target signal. How does it do this?

Say we have measured the Doppler frequency shift due to the speed of a target. We concentrate our receiver on the target frequency and filter[4] out all other frequencies. This can significantly reduce clutter power in the target frequency cell, for the following reason. Most clutter is stationary (such as the ground) or moves only slowly (such as ocean waves or tree branches). So if we are seeking a target that moves much faster than the clutter (a fighter plane over a forest, a torpedo near the sea surface, an insect flying over the ground—recall fig. T12.1) then the clutter will be in a different frequency cell than the target. Problem solved—or greatly ameliorated.

I can illustrate the effects of frequency filtering via a computer simulation. In figure 3.2 you can see a simulated clutter signal, and the same clutter plus a small target. This echo signal is plotted versus time, and you cannot see the target at all, because it is buried beneath clutter. Also shown is the clutter power, and the clutter-plus-target power, in a small frequency

---

[3]We will see in a later chapter that the Doppler effect is described only approximately as a frequency shift. In fact it is a time dilation: the difference between these interpretations is negligible for all radar operation, but for sophisticated acoustic remote sensing as practiced by bats, for example, this subtlety needs to be taken into account.

[4]I will not discuss the mathematical algorithms that are applied to perform these filtering processes. If you intend to delve deeper into remote sensing then you will soon encounter the Fourier transform, which enables us to convert a time signal into a frequency spectrum, or vice versa. Via Fourier transforms we can flip back and forth between these two domains (time and frequency) with ease. A remote sensor receives signals that change with time—they are in the time domain. By transforming to the frequency domain we can see how strongly each frequency is represented in the signal. By chopping out—filtering—frequency components that do not contribute to the target signal, we enhance detection.

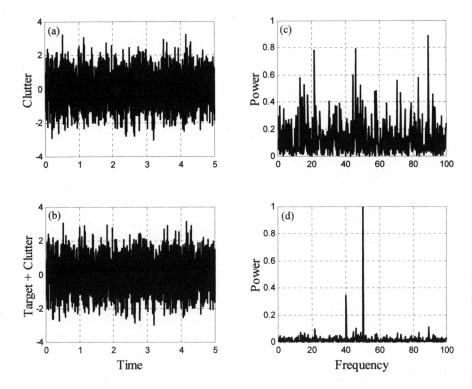

Time   Frequency

Figure 3.2   (a) Simulated clutter. (b) The same clutter plus a small target signal consisting of two tones: one at 50 Hz is 15 dB below the mean clutter level, the other at 40 Hz is 18 dB below. Nothing of this target signal is visible. (c) Clutter power versus frequency—the biggest power sample is here set at 1. Only a small section of the frequency band is shown. (d) Clutter + target power in the same frequency band. Note how both target frequency components stand out—most of the clutter is outside the band and has been rejected.

band of bandwidth 100 Hz around the target frequency. Note how the target stands out. Most of the clutter power is at different frequencies, outside this narrow band, and so has been rejected.

If we make the processed bandwidth even smaller than the 100 Hz shown, then surely we will reject more clutter? Also, recalling that noise power is proportional to bandwidth, we would reject more noise too. True, and the SIR would increase accordingly—a good thing. However, we encounter another trade-off. We cannot be sure exactly what the target speed (and hence Doppler frequency) may be. It can change, as a fighter plane dives or as a moth changes direction, and so we must allow for this variation in target speeds by allowing into our receiver a band of frequencies that covers all likely target speeds. We want our bandwidth to be just wide enough for this purpose, and no wider.[5]

Advanced Doppler processing techniques include an adaptive filter bandwidth: once a target is detected, then the bandwidth narrows, to increase SIR. If the target wants to avoid being tracked, it may change speed or direction—the corresponding change in Doppler frequency must be followed by the processor. Such cat-and-mouse games are elucidated in a later chapter. Doppler processing is also utilized, as we will see in chapter 6, in other advanced processing techniques that classify and even identify the target. To provide a preview of how such target information is attainable, I can extend the simulation of figure 3.2 to include narrower, adaptive filters and show you how the target signal, originally buried under clutter and noise, may be reconstructed. This is shown in figure 3.3. Because my computer is simulating the Doppler processing, we know what the true target signal looks like, and so can compare it with the reconstructed version, obtained by filtering out all the clutter. You can see that it is possible to form a reasonably accurate reconstruction of the target time-domain signal. Starting from this reconstructed signal, further sophisticated processing can help us to classify the target, using techniques discussed in chapter 6.

Suppressing the Clutter

One of the earliest applications of the Doppler effect was to suppress clutter. An airborne radar looking downward will receive echoes from the ground. These ground returns will in general be very strong, because the area of ground that is illuminated by the radar beam at any given instant will be very large. If our plane is a fighter, looking for a bomber target at lower altitude, then the bomber RCS may be much less than the ground clutter RCS. However, the bomber is moving relative to the ground[6] and

---

[5]Some targets are intelligent and know all about the advantages of Doppler processing. They do what they can to prevent us from picking them out in the manner shown in figure 3.2. Thus, a submarine will drift at the ocean wave speed, when its periscope is up, so that our ASW (anti-submarine warfare) radar will have difficulty picking out the periscope. The submarine commander is trying to deny us the advantages accrued from our Doppler processing. Moths are less intelligent than submarine captains, but evolution has led them to avoidance strategies when they detect bat echolocation signals. One strategy is for the moth to go to the ground, or attach itself to a tree branch, and so be at the same speed as the clutter. It then relies on camouflage to remain undetected.

[6]So is the fighter, and this complicates the processing somewhat. However, what matters for clutter canceling is that a difference in speed exists between clutter source and target.

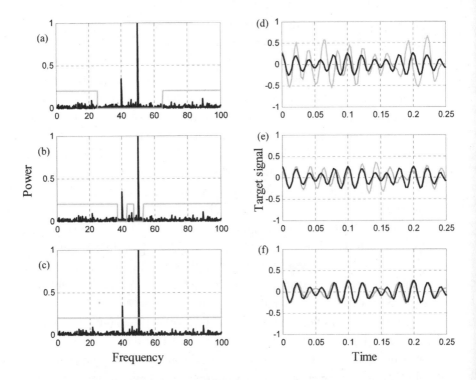

Figure 3.3 *The same filtered target + clutter spectrum as figure 3.2, but with further processing. (a) Narrower filter. (b) Much narrower filter. (c) Target components selected by applying a threshold. (d) Reconstructed target signal (gray) versus original target signal (bold black) for narrower filter—not very good. (e) For much narrower filter—better. (f) Reconstructed from threshold-crossing frequencies—pretty good.*

so the target and clutter will have a different Doppler shift—the target echo will be displaced in frequency. Hence, by looking at the frequency spectrum of the echoes, we can separate moving targets from ground clutter.

The first generation of radars achieved this separation by a simple act of subtraction, known as *clutter canceling*. Imagine a photograph of the scene, taken by the fighter pilot. The bomber is shown below, but it is difficult to pick out against a background of fields and trees and houses. Now imagine the pilot taking another photo just a second later. If the two photos are placed one atop the other, they will match up pretty well, except that the bomber will be displaced, because it has moved in the interval between photos. The stationary ground clutter has not moved. So, subtracting one photo from the other should leave us with two pictures of the bomber, standing out against a much-reduced clutter background. This is what clutter-canceling radars did. One train of radar echoes was subtracted from the next train, leaving a residual echo that originated predominantly from moving targets.

You can see that this might work well for stationary clutter, but also that some difficulties must be overcome before clutter canceling can be workable in practice. First, we have to ensure that the two pictures are aligned properly. That is, the camera must point in the same direction, and this pointing direction must allow for any possible movement of the fighter plane between pictures. Second, the clutter-canceling procedure will not totally remove the clutter if any clutter components are themselves moving. For example, our bomber target was set against a background of fields and trees and houses. Well, the fields and houses may be relied on to remain where they are between pictures, but the trees might sway in the breeze, and so appear different in the second picture. In this case the subtraction process will retain some tree clutter. Of course, our target is moving faster than the trees, and so will have moved further between pictures.

In some circumstances, clutter canceling worked well, but not always, as you may judge from the last paragraph. (It was found to be most appropriate when used in ground-based radars, which are themselves stationary. The clutter of fixed objects such as nearby mountains could be reduced significantly.) The next step in clutter suppression required the next generation of radar signal processors. There was a type of radar, called MTI (for *moving target indicator*), that followed the old clutter-canceling trick with much more sophisticated processing, to provide more effective clutter suppression. MTI processing worked for noncoherent radars, and also for the newer coherent radars. It could be used on moving radar platforms such as airplanes. MTI could be adjusted to suppress some clutter that moved at a specified speed, as well as stationary clutter. (If your radar is aboard an airplane, then the ground is moving, so far as the radar is concerned.)

Modern pulse-Doppler (PD) radar processing techniques have largely, but not entirely, superseded MTI technology. The big difference between Doppler processing and MTI is that Doppler processing is performed in the frequency domain, rather than in the time domain. That is, the processing is performed on the echo frequency spectrum rather than on the echoes themselves. PD radars can estimate the target speed, as well as suppress clutter, and this is something that MTI is not capable of: it merely tells us *that* a target is moving, not how fast. PD can see much smaller targets than can MTI because of coherent integration or video integration of long pulses (see next section). Additionally, modern radar systems in general consist of much more stable and physically robust electronic components—this helps with processing long signals because we do not have to worry about the electronics doing a backflip during the data gathering.

One advantage of MTI is that it works with short bursts of data, rather than with long streams of pulses. This is an advantage in a military context because it means that MTI operation is less vulnerable to enemy *jamming*. Jamming is part of electronic warfare, which I will discuss in more detail in chapter 4. For this reason, MTI processing can be found in modern radar systems as a separate mode of operation. So, a radar operator may briefly switch his radar into MTI mode if he finds that strong clutter is obscuring his targets, while an enemy is trying to jam his signal.

Modern PD radar systems work in both the frequency and time domains. We have seen how, by separating in frequency the radar target from interference, it is possible to boost the target signal. Now I will show you how this can be achieved in the time domain—working directly with the echoes as they are picked up by the receiver. Together, frequency and time domain processing provide powerful tools for detecting weak target signals amid strong noise and clutter.

Boosting the Target Signal

SIR is enhanced by increasing the target signal, as well as by reducing interference signal power. The basic method of achieving this, and the first thought of any signal processing engineer, is *integration*. At its simplest, integration is just adding together the signals from all the pulses in a pulse train. Of the many different integration schemes, I will describe two: *coherent integration* and *video integration*.

If our remote sensor transmits coherent waveforms, then we can improve the SNR by coherent integration. This simply means that we add together a series of consecutive echo pulses, as shown in figure 3.4. Because the echoes are coherent, we know how the phase changes from pulse to pulse. If these pulses reflect off a target and back into our radar receiver, then this phase relationship will be preserved, so long as the target is at a fixed range, or moves only slowly. Knowing the phases, our signal processor can line up all the echoes, as shown, and then add them up to form a bigger signal. Suppose now that an echo signal in the target frequency bin also contains noise. The phase change between consecutive noise echoes is completely random, and so when we try to line them up, the noise signals will still have a random phase, as shown. When added up—integrated—they produce a signal that is much smaller than the integrated target signal. This is because

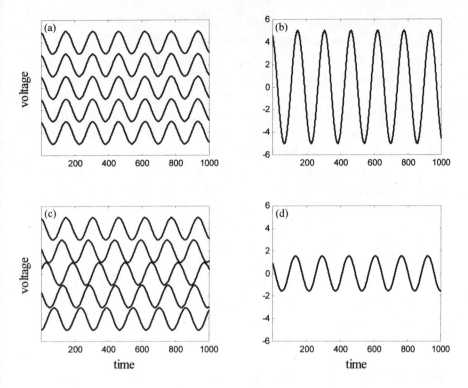

Figure 3.4 *(a) Five target signals phase corrected so that they align. (b) The five signals are added together—coherent integration. (c) Five noise signals in the target frequency bin cannot be aligned. (d) So, they add up to make a smaller signal.*

the random phases mean that the wave peaks and troughs do not align, and so when summed the echo signals can just as easily cancel each other as reinforce each other. It is like stacking bricks (fig. 3.5). If the bricks are carefully stacked one atop the other, they reach a greater height than if they were just thrown together in a pile. We expect that the pile of bricks will be higher than a single brick, but not nearly as high as when bricks are neatly lined up. We know from theory what the improvement in SNR will be, due to integration. If N pulses are integrated, then the ratio of target signal power to noise signal power will increase by a factor N, on average. (Because both target and noise signals fluctuate, all our results are statistical—hence the "on average.")

This type of integration is the most efficient possible—no other integration scheme can produce an *integration gain*, as we call it, that exceeds N, even in theory. In practice the benefit of integration is not usually so good.

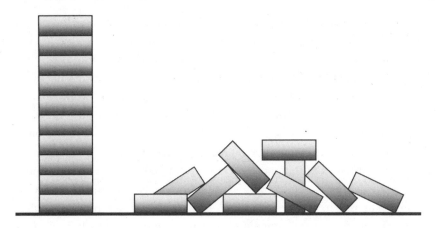

Figure 3.5 *Coherent integration of bricks (left): when they are aligned, they add up to a greater height than if they are just piled together randomly.*

For sure, if we have a stationary target with an echo that is buried in electronic noise, then we can coherently integrate N echo pulse and increase the SNR by a factor of N. But there are limitations. First, real-world targets of interest often move. Because they move, the target range changes a little, from pulse to pulse. A small change in target range can mean a big change in phase between consecutive pulses, and so we lose the ability to line up the echo signals prior to integration. In practice this limits the number of pulses that we can usefully integrate. Put another way, the amount of time that our radar or sonar system can stare at a target is limited, because sooner or later target motion will cause the phase of echo pulses to *decorrelate*. This decorrelation arises due to target rotation (unless the target is a simple sphere) and target speed—any kind of movement will sooner or later mean that we lose the ability to line up target echo phases. Target correlation time is an important parameter in detection theory. A typical complex radar target may have a correlation time of 100 ms. (Sonar correlation times, and integration times, can be much longer.) This means that we can coherently integrate a train of echo pulses that is less than 100 ms long. If we attempt to integrate a much longer pulse train, we get no further benefits. In fact, the SNR will fall, because we are effectively just adding in more noise.

A second limitation imposed on coherent integration is that of clutter correlation. So far I have integrated a target signal in noise, not clutter, because this is the most clear-cut case for demonstrating the benefits of integration, in general, and of coherent integration, in particular. However, a lot of the interference that a target echo must compete with is due to clutter, rather than to noise. Clutter is much more varied than noise, as we have seen, and, in particular, it can be correlated. After all, if part of a sonar pulse is reflected off a rock, then how will the sonar signal processor know that the echo pulse came from a rock and not from a target? If both rock and target are in the same resolution cell, then there is no way of knowing—both rock and target components of the echo signal will be integrated. Some clutter sources—like stationary rocks—have a very long correlation time, whereas other types of clutter—say a kelp frond that is attached to the rock, and is fluttering in the ocean current—have short correlation times. Clutter is so varied that in many situations the clutter echoes will be from sources with long, medium, and short correlation times. So what happens to the signal-to-clutter (SCR) ratio when we integrate a train of echo pulses that came from a resolution cell containing both target and clutter? Well, it will increase, but not as much as the SNR. If we coherently integrate N pulses then the SCR will increase by $N/3$, or $N/2$, or some other amount that is less than N. In other words, integrating a "target + clutter" signal is beneficial, but not so much as integrating a "target + noise" signal. Clutter with a correlation time that is shorter than the target correlation time will add up inefficiently—a pile of bricks—whereas clutter with a comparable or longer correlation time will stack up just as efficiently as the target. So, how well coherent integration performs depends on the composition of clutter and upon the target correlation time.

"Video integration" is the name usually applied by radar engineers, for historical reasons,[7] to *noncoherent integration*. It is used far more commonly than coherent integration is used, despite the fact that it is not so efficient. The early radar systems were not coherent, and so coherent integration was a nonstarter. Instead of adding up the echo signals, integration was achieved by adding up the signal envelopes, or the power levels. You may recall from

---

[7]Old-fashioned radar screens were not the fancy, modern multicolored displays, but instead were primitive phosphor tubes, like the old oscilloscopes that you may recall from school days (those of you who are at least as old as I am). The CH radar displays were of this type. Targets appeared on these displays as dots of light, which faded away slowly. No integration was performed in the signal processor—such processing barely existed. Instead, whenever a target was detected, a dot was placed on the screen, and if the same target was detected again, then another dot appeared. So a stationary target would build up in light intensity, as dots were piled up one upon another. Such targets would stand out from the background noise and clutter. This is why noncoherent integration is still called "video integration" by radar engineers.

note T2 that the envelope is the outline or amplitude of a signal, and this measures the amount of power it contains but not the wave phase. So video integration adds together the power of a pulse train without considering the phase relationship between pulses.

This method of integration works, in that targets become more detectable, but it is not as good as coherent integration, which takes into account *all* the information that is contained in the pulse, rather than just the amplitude information. Video integration is much easier to implement, however, because there is no lining-up process—you simply throw all the potatoes into one pot and see how full it gets. Because it is so easy to do, most radars operating in the world today still employ video integration, even though many of them are coherent. The relative effectiveness of video integration depends on the initial SNR of the target. (Again for simplicity I will assume that the interference is due to noise—if clutter is present, then the situation becomes more complicated, as for the coherent case.) If the target power is initially quite large compared with noise power—high SNR—then video integration can be quite efficient. If the SNR is low, then video integration is inefficient. This is the wrong way round—we would like video integration to be more efficient when SNR is low. But life is like that sometimes: the more you have, the more you get. The radar text books to which I and my radar colleagues continually referred, when we were trying to estimate the performance of our radar designs, contain pages upon pages of graphs indicating the efficiency or inefficiency of video integration, depending on SNR, target statistics, and the number of pulses integrated. If the interference is dominated by noise, then many of these curves can be calculated from the principles of detection theory. If clutter is present, however, such calculations become intractable.[8]

Skilled and experienced humans have an ability to integrate radar and sonar signals, via some unconscious brain function (something to do with pattern recognition). *How* we do it may be poorly understood, but *how well* we do it is known, and has been studied extensively. A modern sonar operator can integrate complex signals and visually spot a target on a display that looks just like noise to you and me. After an hour or so the operator becomes less efficient due to fatigue. Such human abilities were first noted during the Battle of Britain when radar targets at extreme range,

---

[8]Because of this intractability, these days, radar systems engineers resort to mathematical modeling and computer simulation to estimate the effectiveness of integration, when clutter is present in the echo signals.

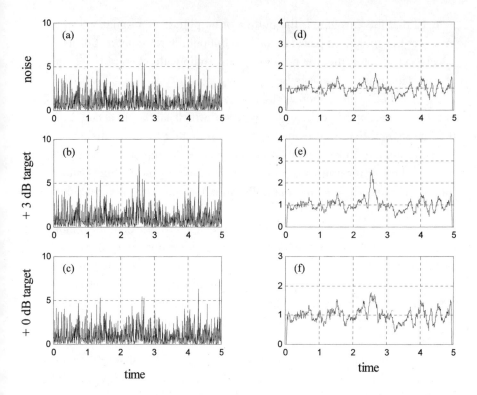

Figure 3.6 *Simulation of video integration. (a) 100,000 samples of noise power, fluctuating in time. (b) The same noise signal, but with a 3-dB target added, at time 2.5. (c) This time the added target is smaller, at 0 dB. (d) Noise power integrated over 20 samples—note the reduced fluctuations. (e) Noise + 3-dB target, integrated over 20 samples—the target stands out. (f) Noise + 0-dB target, integrated over 20 samples— the target can be seen, but it only just exceeds the noise. Video integration does not increase the SNR, but it does improve target detectability, by reducing fluctuations.*

with echo signals well below noise power levels, could be detected by WAAF operators.

I can provide you with an indication of the way in which video integration works by again appealing to a computer simulation. In figure 3.6 you can see that video integration of a signal that includes a weak target makes the target more obvious. We say that the target has become more detectable. Careful perusal of the figure will also show you, rather puzzlingly, that the SNR has not increased. Rather, the fluctuations have been damped down as a result of integration. This apparent conundrum—how target detectability increases without improving the SNR—is explained in note T16.

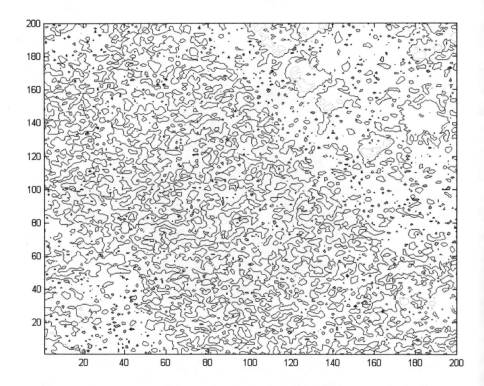

Figure 3.7 *Map of real radar ground-clutter data. The clutter power is represented by contours. Note how the clutter varies over the map: echo power varies with location and over different length scales, due to the underlying topographical features.*

As you might expect, the explanation lies in the statistical nature of echo pulses.

Constant False Alarm Rate (CFAR) Processing

So, we have suppressed clutter and integrated the echoes to boost target detectability. Now we need to set a threshold—draw a line in the sand—so that we can say to any echo that crosses it "you're a target." How do we know where to set the threshold power? From detailed calculations, given an acceptable false alarm rate, a desired detection probability, and given some assumptions about clutter statistics, we can determine how high (relative to the mean clutter level) to set our threshold. The difficulty is that the mean clutter level changes in time and with location.

This point is illustrated in figure 3.7. In this figure I have taken some real radar data of surface clutter—a mixture of buildings and natural topographical features, as seen by an airborne radar system—and plotted a

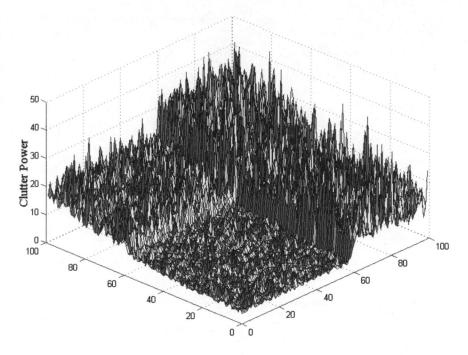

Figure 3.8 *Simulated clutter power versus position on the ground. Note how the mean power level is changing from place to place. In the foreground, clutter power fluctuates about a relatively low mean level, whereas in the background the mean level is high. It makes little sense to refer a radar detection level to the overall, global mean power level—instead, the threshold should be set relative to local clutter power.*

contour map of clutter power. Different features of the landscape contribute differently to the clutter power, and you can readily appreciate from the figure how *variable* the clutter level can be. The power changes with location, and it changes more rapidly in some areas of the map than in others. I can better show the problems that are created by this clutter variability by resorting to simulated clutter data, which I can construct in such a way as to bring out the main point without also displaying a myriad of other features that are inevitably to be found in real data. So, please consider figure 3.8. Here you can see how clutter power changes with location, sometimes quite quickly and substantially. Now imagine trying to set a threshold power level so that we can see whether any targets are hidden in these data. The calculated threshold is, let us say, 30 dB above the mean clutter power. But from figure 3.8 it is clear that it makes little sense to work with the mean power of the entire scene, since this varies so much. Were I to persist with this strategy and set a global threshold, based upon the global clutter average, then I would obtain a lot of false alarms from the background part of the figure, since the local average clutter power level here is higher than elsewhere. Similarly, I would likely miss some targets that are in the foreground, where the local clutter level is lower—I could reduce the threshold in this region. In short, by setting a global threshold I would find that the false-alarm rate would vary from one location to the next.

It would be much better if the false alarms were evenly spread over the map, so that no local clustering of them would deceive us into thinking that they are targets. As a consequence, we need to define *local* clutter as mean power levels, so we know where to set the *local* threshold. This task is performed by a CFAR (constant false alarm rate) processor. In note T17 you will see how CFARs work, and the benefits of CFAR processing for target detection.

Improving Target Angle Estimates

At this point in the book, we have a radar or sonar system that can:

- transmit a train of pulses
- receive the faint echoes from targets
- boost these echoes so that they stand out against the interference background
- threshold-test echoes to declare targets
- estimate target position, to within a resolution cell, and target speed.

But we want more. The resolution cells can be enormous, and we would like to be able to specify more precisely where the target is to be found at a given instant. Such improved accuracy will help to improve the SIR (recall that most of the interference that competes with a target signal comes from the same resolution cell as the target[9]). Improved accuracy will help us, in a military context, to direct fire at the target. A search-and-rescue mission would like precise information about the location of capsized boats, and so on, so that their helicopters do not waste time scouring the ocean surface. Sonars that are searching for mines near a harbor entrance, or

---

[9]In fact, some interference can appear from outside the target resolution cell. Later we will see that aliasing allows clutter to foldover from longer ranges into the target range cell. Also, the antenna sidelobes permit nearby clutter to enter into the receiver even though the mainlobe is pointing in a very different direction. This sidelobe clutter is greatly suppressed because sidelobe gain is much reduced compared with mainlobe gain, but a nearby strong clutter source may break through.

sunken ships on the ocean bed, will be more effective if they are more accurate.

In the next few sections I will discuss some of the methods whereby the resolution cell size of a radar or sonar system can be reduced. The "natural" cell size is determined by pulse duration (in the case of range) and by mainlobe beamwidth (in the case of azimuth and elevation angles). We can improve on this natural limitation to position estimation accuracy by implementing some clever signal processing. First, I will look at how to improve angle estimation.

We have already seen one example of improved angle estimation: the amplitude monopulse technique (note T6) utilized by the old CH radar system in the Battle of Britain. This method works well if the radar system is capable of producing two beams simultaneously. Most radar systems, especially portable ones, do not have this luxury—the transmitter antenna produces a single beam that is scanned mechanically in azimuth and elevation, and likewise for the receiver antenna.[10] For such systems we need a different method, and a *sector scan* will do the trick for us. Not only does this scheme improve angle estimation accuracy, but it also improves angular resolution.

Consider search radar, say a land-based system that is tasked with searching the skies just above the horizon for incoming aircraft or missiles. It scans the horizon in just the way that a lighthouse beam does; by rotating at a constant rate. Instead of sending out light, our radar emits microwave pulses and receives the echoes. We may regard the transmitter and receiver beam as covering the same area at the same time (it would be useless if high-gain transmit-and-receive antennas pointed in very different directions, since the target echoes would not be picked up by the receiver). Our search radar *scan rate* is determined as yet another trade-off. Let us say that the azimuth width of the radar beams—both transmit and receive—are 3°. To intercept a target that approaches rapidly from over the horizon, we would like the antenna to rotate very fast. This is so that we can detect our target at the earliest possible moment. For example, if the scan rate were slow, say one complete rotation per minute, then a fast incoming missile might reach us before our radar beam got pointed in the right direction to detect it. So, faster is better.

---

[10]In chapter 6 we will see that the next generation of phased array radars will be capable of producing many beams at once from the same array. The current generation of phased array radars is able to steer a beam electronically, without physically moving the array.

But only up to a point. If the scan rate is too fast, then we may not detect a target even if we do illuminate it. This is because our 3° beam will sweep past the target quickly and so not many of our transmitter pulses will intercept the target. To give a concrete example—suppose that the scan rate is 30° per second, so that one complete rotation takes 12 seconds. Because our beamwidth is 3°, the beam will illuminate a target for only 0.1 seconds. If pulses are transmitted at the rate of 1,000 s^{-1} (1 kHz), then 100 pulses will echo from the target each time our radar beam sweeps by. Now we decide that one look every 12 seconds is too slow an *update rate* for our search radar—we would like to see targets more frequently than this, to establish their heading direction and to confirm that they are what we think they are. We increase scan rate to 180° s^{-1} so that the targets are illuminated once every two seconds. Now the number of pulses that illuminate the target per scan (the number of *hits in the beam*, in radar parlance) drops from 100 to 16 or 17. We would like the number of hits to be as large as possible so that integration permits us to boost the target signal and detect the target. Clearly it is not possible to scan very fast and also maintain a high number of hits.[11]

In practice our search radar will hit the target with a limited number of pulses per scan. The *sector scan* method can improve accuracy based on the known number of hits in the beam. This also applies for tracking radars. With trackers the radar transmitter beam does not sweep through an entire circle at a uniform rate, rather it scans back and forth across a limited sector where a target is known to be. The same sector scan scheme applies, but the increased update rate helps with tracking the target. Hence, the name of this technique, which is commonly used by tracking radars, is *sector scan*.

Sector Scan Technique

The basic idea can be grasped from figure 3.9. In the case illustrated, our radar has five hits in the beam—five pulses illuminate the target during the interval (the *dwell time*) during which the beam sweeps past. The power of each pulse that echoes from a target will not be the same; it will vary because of the transmitter beamshape, as illustrated in figure 3.9. If we can

---

[11]You might think that we could simply increase the pulse repetition frequency (PRF), so that microwave pulses spew out of our transmitter faster. This would allow a higher scan rate while maintaining the same number of hits on the target. Unfortunately, limitations are placed on the PRF by the microwave power source. Also, very high PRFs would require shortening the duration of each pulse, which reduces its power and so reduces detection range. Choosing radar parameters is all about trade-offs between different competing performance requirements.

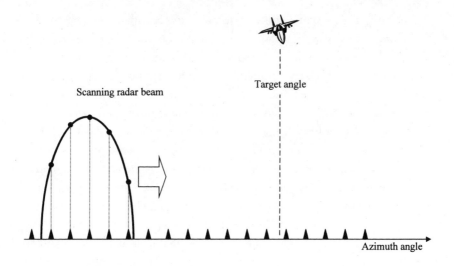

Figure 3.9 *For a constant scan rate, the transmitted pulses (markers) occur at equal angle intervals. As the radar main beam scans past a target, the target is illuminated by pulses with different power, depending on where in the beam the pulse falls. Knowing the beamshape and the number of hits in the beam (here 5) we can apply a sector scan algorithm to refine target angle estimate.*

assume that the target RCS does not change during the brief interval that it is in the beam (a reasonable assumption, but by no means a dead certainty), then we would expect the echoes from the target to have the same varying power. Now, the essence of the sector scan technique is this: we *know* the beamshape and so, from the way in which our target echoes vary in power, we can estimate where the target is located. You can see from figure 3.9 that we can fit our beamshape to the five target echoes, and so obtain a more accurate estimate of target location. That is, we can estimate target azimuth angle with an accuracy that is better than the antenna beamwidth.

There are a number of sector scan algorithms, and some work better than others in different situations. If our antenna scan rate, pulse repetition frequency, and beamwidth are such that there is only one hit in the beam, then our best estimate of target angle is simply the angle at which the echo appears—no surprises here. The estimation accuracy in this case is simply the beamwidth, say $\Delta\theta = 3°$ in this case. With more than one hit in the beam, sector scan comes into its own. The so-called centroid algorithm says: "pick the middle pulse." So we estimate the target angle to be that which corresponds to the middle of our five target echoes. This simple algorithm improves angle estimate accuracy by a factor of about 5, or n for n hits in the beam. So in our case the target bearing angle will be estimated to within $^3/_5$ of a degree. Another sector scan algorithm is a little more sophisticated. The "center-of-mass" algorithm weights each echo according to its power, to attain an angle estimate that is better than $\Delta\theta/n$. Both these algorithms are limited in ultimate accuracy by noise and target RCS power fluctuations, so that in practice we cannot improve accuracy, by increasing n beyond $n = 10$ or so. Another variant of the sector scan technique is the "beamshape match" algorithm that matches the echo power variation with the known beamshape. I show a simulation of this technique in note T18, where you will also see how angle resolution, as well as accuracy, is improved. The beamshape match approach is our first example of *correlation*— comparison of an echo pattern with a known shape, in this case beamshape—of which more soon. Beamshape match improves accuracy to about $\Delta\theta/25$ for large n. So, for our $3°$ beamwidth we may be able to obtain an azimuth accuracy of $0.12°$, say an eighth of a degree—pretty good. The beamshape algorithm is the most efficient of the sector scan techniques at high SNR (see note T18), but for intermediate SNR the center-of-mass algorithm has been found to work best.

Sector scan techniques may be applied in both angle directions—elevation as well as azimuth. This is appropriate for a tracking radar, but not for the horizon-search radar we considered previously. To improve both angle estimates, the tracking radar beam must scan around the target in both azimuth and elevation. This is usually accomplished via a *conical scan* that, as the name implies, involves rotating the beam in a cone that circles the target direction. Sector scan algorithms are simple to implement for modern radar systems, which have signal processors with huge number-crunching capabilities, so that the rapid real-time calculations required for sector scan angle estimate in two dimensions are performed with ease.

Improving Target Range Estimates

Here is where I get to tell you about two very important techniques. *Frequency modulation* (FM) is widely used in remote sensing applications to improve both range accuracy and range resolution. Before I can tell you about FM, though, you need to know about signal correlation. Correlation is an important concept that goes beyond improving target position estimates; it has wider application to many branches of signal processing, and beyond signal processing, it is a statistical technique that is much used. Here I will restrict attention to the correlation of signals, because this aspect is most relevant to our story and because it can most readily convey the idea behind correlations with a few simple pictures, rather than pages of math. The Masters of the (remote sensing) Universe, that is, bats, almost certainly utilize correlation processing, and that's good enough for me.

Correlation is not a property of a single signal, but rather it determines the similarity of two signals. The *correlation function* of two signals is a quantitative measure of how similar they are. It tells us, typically, that signal A did not look much like signal B a short while ago, but right now the two signals exhibit a passing resemblance, though in an instant this similarity will disappear. We can see how such a function is going to be extremely useful to us. Because an echo is a (mangled) version of a *known* transmitted pulse, we can compare this known pulse with the stream of receiver signals. This comparison will assist us in picking out the echoes from all the other EM/acoustic garbage that is poured into our receiver (from space, from other radar systems, from communications masts . . . from the ocean, from other sonar systems, from ship engine noises a hundred miles away, from copulating shrimps . . .). It is like a border guard who is looking for a spy among a stream of people passing through a turnstile by comparing the faces he sees with a photograph of the spy.

In note T19 I provide a simple pictorial explanation behind signal correlation—simple yet directly relevant. From this it is clear, I hope, how correlating an echo pulse with a copy of the transmitted pulse enables us to improve target range accuracy and resolution. So, instead of the "natural" range accuracy of $\Delta R = \frac{1}{2} c\tau$ we can obtain range estimates that are a small fraction of ΔR.

We turn to FM ranging to drastically (and I mean *drastically*) improve range estimation accuracy and for discriminating multiple targets[12] that are separated in range by less than ΔR.

Chirp FM Ranging

No, I am not returning to a digression about our feathered friends. The term *chirp* is a technical one, for reasons that will soon be clear.

FM waveforms change frequency as they unfold. The linear-FM waveforms considered here increase (or decrease—it doesn't matter) frequency

---

[12]Such as a herd of German bombers heading toward the coast of southern England. Thankfully, modern Germans seem disinclined to pursue such activities. I am sure that Air Marshall Dowding and Fighter Command would have greatly appreciated FM ranging had the technique been available at the time since, you may recall, they had some difficulty in estimating the strength of enemy raids. No such problems exist with modern radar systems.

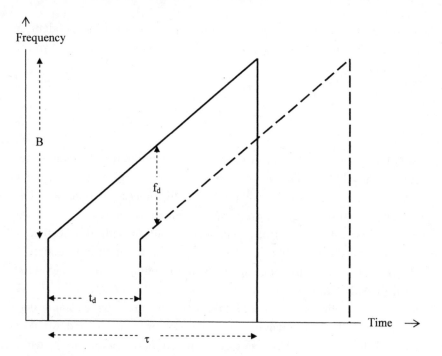

Frequency

B

f_d

t_d

τ

Time →

Figure 3.10 *Linear FM ranging. A transmitted pulse (solid line) increases in frequency by an amount B over the pulse duration τ. The echo pulse (dotted line) arrives back a time t_d after transmission. It has the same chirp waveform, but at any given instant the frequency is less than that of the transmitted pulse by an amount f_d. Measuring f_d gives us t_d and, hence, target range.*

proportionally with time, as shown in figure 3.10. The starting frequency is not important to us; what matters is the pulse bandwidth B—this is the amount by which the frequency of the pulse changes from start to finish. Suppose that a pulse with this waveform is transmitted by a radar and that· an echo is received from a stationary target. The echo is also shown in figure 3.10. It returns to the receiver a time t_d (d for delay) after being transmitted. From this delay time we know that the range of the target is about $R = \frac{1}{2}ct_d$, but we also know that the error of this estimate is $\Delta R = \frac{1}{2}c\tau$ where τ is the pulse length. But look at figure 3.10—here the delay time is less than the pulse length. The essential point is that, because of the linear-FM waveform, the echo pulse and transmitted pulse differ in frequency by a constant amount f_d during the time when the two pulses overlap. Now f_d is proportional to the delay time t_d, and so if we can measure f_d then we can estimate target range. This range estimate, obtained by measuring a frequency, is not subject to pulse length inaccuracy—it is independent of τ. In fact, it can be shown that the error in estimating range for a typical linear-FM ranging system is

$$\Delta R = \delta R = \frac{c}{2B} \tag{2}$$

Note that I have written this error as range resolution as well as range accuracy—it is both.[13] The accuracy with which we can estimate a target range by this method is $c/2B$, and the minimum target separation for which we can discern two targets is $c/2B$. So, by estimating range in this way we can make the estimation error independent of pulse length. If we can construct a pulse waveform that has very high bandwidth, then from equation (2) you can see that the range accuracy and resolution can be made very small indeed. The ratio of "natural" accuracy $\Delta R = \frac{1}{2}c\tau$ to the linear-FM accuracy—equation (2)—is $B\tau$, and this is known as the *pulse compression factor*. This factor is well named since, by adoption of this method, we have been able to attain the range accuracy of a much shorter pulse, though without losing pulse power by actually reducing its duration. In practice pulse compression factors of 50 or 100 are common in the radar field, so that target ranges can be determined with great precision.

I hope you realize that linear-FM signal processing requires a coherent radar (or sonar), since we need to take into account the changing phase of the pulse waveform. In fact we can perform the range estimation within the time domain, rather than by measuring the frequency f_d, by making use of correlation processing. Just such correlation processing of a linear-FM waveform is carried out in note T20, via another computer simulation. I will take the opportunity in note T20 to add noise into the echo waveform, and show you that this makes no difference to the ability of the correlator to es-

---

[13]Some of my erstwhile colleagues in the radar systems field might object, rather pedantically, to the manner in which I have explained linear-FM ranging. For example, I suggested that an echo might return before the transmitter has finished sending the pulse. This is technically possible, but in practice such an echo would not be detected because of *transmitter blanking*. Because radar receiver antennas are usually sitting right next to the transmitters, it is necessary to ensure that the transmitter is shut off when the receiver is listening—recall from the range equation how small the received signal can be. Without blanking, the receiver would be blinded (or deafened, if you prefer the analogy made in the Introduction) by the transmitter and would detect nothing. To appease my colleagues, you might want to think of the echo pulse (from whatever range) as being compared with a copy of the transmitter pulse—they are being correlated—to determine the echo time delay with sharpened precision.

timate time delay—and hence range—with the precision of equation (2). Also, you will see explicitly the pulse waveform shown schematically in figure 3.10, and how this pulse is time compressed.

The linear-FM waveform is called a chirp waveform because, if it were an audio signal, it would rise (or fall) in pitch, and sound a bit like a bird chirping. In fact, though, sonar systems (and echolocators) do not make use of linear FM as often as radar systems do. This is because, for acoustic application, there are issues with the chirp pulse waveform that do not apply for EM sensors. We will examine this problem, and how bats and sonar systems engineers overcome it, in a later chapter. For now, we note that there is another serious problem with our chirp processing: what if the target is moving? A moving target Doppler-shifts the echo frequency, which adds to f_d. So a moving target, if not identified as moving, will lead to an erroneous range estimate by chirp processing. The problem is serious but not a show stopper. If we know that the target is moving then we can compensate for the movement. Other processing tricks exist to overcome this problem, but the best method is that adopted by bats. To be continued.

A common thread runs through all the techniques that I have described in this chapter for improving our knowledge of target position. We have sharpened up angle estimation by sector scanning, and improved range estimation by correlation of transmitted and received signals. In both cases we are applying extra information that we know about our remote sensing system—information that was not utilized to form the "natural" resolution cell. In angle estimation, we were making use of our knowledge of antenna beamshape. For improving range estimates we coded the transmitter pulses in some way—by Barker codes or linear frequency change—and looked for the same code in the receiver signal.

4 ....

Tactics: Skunks & Old Crows

If chapter 1 consisted of an historical aperitif, then appetizers and entrées were served in chapters 2 and 3. Desserts, digestif, port, and cigars are yet to appear. That is to say, what follows will contain less meat and potatoes than the last two chapters but will, perhaps, be easier to digest. I do not mean to imply that the content henceforth will be nontechnical, but that in general it will be less mathematical and more descriptive. You are halfway through the chapters, but have encountered 20 of the 23 technical notes. You now have the tools that permit me this change of emphasis; I can describe a signal processing goal or technique (angular resolution, say, or Doppler processing) without having to pause and explain.

We now have a highly capable remote sensing system. Our system can transmit complex pulses and dig out their echoes amid noise and clutter. We can set a local threshold over the scene that we are sensing, so that a known constant false alarm rate results, when we declare an echo to be a target. The target position has been estimated with high precision and, if it is moving, we know how fast. We can track the target to ascertain its heading. The impressive capability of radar and sonar systems has evolved since World War II, frenetically at first and rapidly thereafter. In a military context, the potency of radar systems has posed significant and increasing threats to the radar targets, be they on land or sea or in the air. However, these targets have not stagnated during the 70 years since radar first burst onto the military stage. They have themselves developed and have learned to deceive our remote sensing equipment—deceive and strike back. In this chapter I will discuss this evolutionary combat between hunter and prey, initially from the perspective of a radar systems engineer and then with significant examples drawn from the sonar world.

For radar a large measure of this "evolutionary combat"—the running dogfight between detector and target—comes under the heading of electronic warfare (EW) with subheadings electronic countermeasures (ECM) and electronic counter-countermeasures (ECCM). It is a story of leapfrogging arms races, of cloak-and-dagger deception, fought for possession and control of electromagnetic territory—the EM spectrum. I will describe this ethereal landscape, or at least the province that contains our radar systems, and then show how it has been fought over during the past seven decades. Radar targets seek to hide behind electronic camouflage. They try by subterfuge to turn the tables on the radar systems that seek them out, and these radar systems respond to parry the threat. I will describe the methods developed over the years by both predators and prey, and will then summarize the ECM and ECCM shenanigans that took place over the skies of Europe during World War II.

The title of this chapter comes from two groups of people who can be taken as representative of the radar predators (sometimes prey) and the target prey (sometimes predators). I do not propose to describe to you the history of the Association of Old Crows or of the Lockheed Skunk Works, or to weigh up their contributions to EW. Instead, please accept these two organizations as icons— representatives of the United States in the EW world. The Association of Old Crows (AOC) is an "international professional association engaged in the science and practice of electronic warfare," according to its Web site. It represents the thousands of electronic engineers (largely American) who have spent their careers trying to outfox the other side: radar engineers craning to see past the wall of smoke-and-mirrors that has been set up by, well, other AOC members. The Lockheed Skunk Works stands for the manufacturers who develop the hardware to implement this electromagnetic warfare. In fact Lockheed Martin makes military aircraft, and has done so since before World War II; their Skunk Works (based originally in Burbank, California) evolved out of that conflict and is the droll moniker for Lockheed's advanced development projects—their experimental aircraft. Many of these aircraft never get off the ground, or past the prototype stage, because they are so off-the-wall—creatures of the fertile imagination of a talented bunch of engineers and technicians. They began in

1943 under the protective management of Kelly Johnson, who set up his new group in their own premises, and with a semiautonomous culture geared to concentrate the engineers' minds on technical issues rather than budget constraints. Today, I suppose that this type of setup would be called a research institute, but in 1943 it was all about developing new planes to win the war. Their first success was the XP-80, an experimental jet fighter built around the Whittle jet engine—one of the fruits of the Tizard mission in 1940. Later, as the P-80, this plane became the first U.S. production jet airplane.

The U-2 spy plane was a child of the Skunk Works and, 40 years on, so is the F/A-22 Raptor stealth fighter. These and other aircraft were designed to beat the enemy EW capabilities. The enemies—be they German, Japanese, Russian, Chinese, Vietnamese, or Iraqi—had their own Electronic Warriors and Skunk Works, and so battle was joined and continues to this day. An aviation journalist wrote the plain truth nearly 20 years ago[1] when he said: "Born in the darkest hours of the Blitz, weaned during the night bombing offensive and matured in the skies above Hanoi, Electronic Warfare has become a military discipline within itself with a pervasive influence upon the strategy, tactics and technology of modern warfare."

Radar Real Estate

As with all desirable land, the radar part of the EM spectrum has been carved up into lots, argued over and subdivided many ways. As an indication of this, please consider Table 1, where you will find the letter nomenclature and nominal operating frequency bands of the different breeds of radar systems. Had this book been a textbook for practicing radar engineers or graduate EE students, I would have included a more detailed version at the beginning, for easy reference so that the reader could memorize the letters and their meanings. When radar design engineers talk among themselves they generally do not refer to specific operating frequencies, except perhaps in a detailed project meeting. More generally the speaker will refer to, for example, a "Ku-band system" and this will immediately bring to the listener's mind a short-wavelength, short-range tracker or homing system, probably airborne. At international conferences or in the discussion of concepts or new projects with a competitor, this nomenclature avoids awkwardness, quite apart from being a useful insider's shorthand. The alterna-

---

[1]Carlo Kopp, writing in *Australian Aviation*, September 1988.

Table 1. Radar bands and their main applications

Band designation	Nominal frequency range	Radar application
HF	3–30 MHz	Long-range but low-resolution search radar. CH operated in this band. More modern versions operate over the horizon.
VHF	30–300 MHz	Long-range line-of-sight radars, insensitive to
UHF	300–1000 MHz	weather clutter, low to medium resolutions.
L	1–2 GHz	Long-range search radars with medium resolution but some sensitivity to weather clutter.
S	2–4 GHz	Medium-range search or long-range tracking radars of medium resolution and accuracy, affected by weather.
C	4–8 GHz	Medium-range search or long-range tracking/guidance radars with high accuracy in fairly clear weather.
X	8–12 GHz	Short-range search or medium-range tracker radars with high accuracy and resolution in clear weather.
K_u	12–18 GHz	Small, short-range tracking and guidance radars,
K	18–27 GHz	usually associated with missiles. Vulnerable to
K_a	27–40 GHz	weather clutter.
V	40–75 GHz	Very-short-range tracking radars.
W	75–110 GHz	Very-short-range tracking and guidance radars.
mm	110–300 GHz	As for V, W only more so, with increased accuracy.

tive would be to discuss detailed frequencies, and such information is highly classified precisely because of EW, as will shortly become abundantly clear.

The purpose of my showing you this information is simply to get across the idea that radar has developed and matured, and consequently it has specialized and fragmented over the decades. This specialization is in large part due to the demands placed on radar systems by their targets. Perhaps the targets have such low RCS that the radars must have very small resolution cells and high signal processing capabilities to dig tiny target signals out of clutter. Or else the targets are actively trying to spoof the radars, which have adapted to parry such thrusts. Note that no such specialization of acoustic bandwidth occurs—or almost none. This is not, repeat not, because the chicanery of EW is absent in the underwater sonar world, but rather it is a consequence of the very different medium. Underwater sonar transmissions in general do not travel as far as EM waves do through the air, and so there is much less opportunity for an enemy to eavesdrop or to actively deceive. The aquatic smoke-and-mirrors game is played with different rules, based largely on passive remote sensing. Passive sensing plays a much more significant role in sonar than in radar, and leads to its own, very different, but equally competitive war of wits between sensor and sensed, as we will see.

Window, Rope, Chaff, and Other Decoys

Jam. To a radar engineer this is not what you spread on toast and has nothing to do with Jimi Hendrix: to jam an EM signal is to deny its use to your enemy. Radio signals and radar transmissions can be jammed, but only the latter is relevant to us here. Jamming is a big subject, and has been since the very beginnings of radar, because of the early appreciation of just how important radar is to military effectiveness. Jamming can be divided into two broad categories: active and passive. Active jamming means that you send out your own EM signals to blind or delude your enemy; passive jamming means that you try to block his radar transmissions from giving away your precise location. Let's begin with passive jamming.

Chaff was invented independently by both sides during the air war over Europe during World War II. Both sides were reluctant to deploy chaff, at first, because they thought it would indicate to the enemy that their own radars would be vulnerable to jamming by this technique. In those days chaff consisted of thin aluminum strips, cut to a precise length, and distributed from aircraft. As the small bundles hit the airflow, they burst open to produce a cloud of silvery flakes, tumbling randomly and falling slowly through the air. Originally, the chaff strips were cut by hand, and were dispersed out of aircraft by crew members physically throwing the bundles through open hatches. Nowadays the cutting and dispersal are both done automatically, without the aircraft crew even having to note that there is a need for chaff dispersal. But I am getting ahead of myself.

Chaff is cut to a precise length to foil (no pun intended) an investigating radar that is trying to detect the airplane—typically a bomber. The air crew knows that they have been spotted by the radar and are doing their utmost to prevent it from directing AAA fire or interceptor fighter planes at them. In World War II the AAA consisted of flak artillery, which the Germans deployed massively as Allied air raids devastated their cities.

(Nowadays AAA is a surface-to-air missile or a Stinger, but the idea is the same.) Chaff blocks the radar view, showing up as bright splodges over a section of the radar screen where, just a moment earlier, there had been a formation of enemy planes. The splodges cover the aircraft, denying gun-laying accuracy or even interception coordinates for fighters.

Chaff is cut to a precise length because the chaff strips have a maximum RCS when they are half as long as the radar wavelength. They work for wavelengths that vary by a few percent either way, but on the whole they are effective screens only for a narrow radar bandwidth. For this reason, the operating wavelength of the radars must be known in advance, or else the airplane must store chaff that is precut to many different wavelengths, and the right bundle is deployed when necessary.

Theory predicts that randomly oriented strips of chaff will have an average RCS of $\sigma = 0.18N_c\lambda^2$. N_c is the number of chaff strips that are deployed and illuminated by the radar beam, and λ is the radar wavelength. So, chaff is most effective against long-wavelength radars. Given the huge number of chaff strips that are deployed, this can result in an enormous chaff cloud RCS that blots out everything else in the vicinity. The first use of chaff by the British, in a massive bombing raid against Hamburg in 1943, deployed 92 million chaff strips. In those days, chaff was known in Britain as *window*, perhaps because it provided a window of opportunity for their bombers to hit a German city while remaining invisible to enemy radar. The chaff would eventually disperse or fall below bombing altitude, and so the window would close. This first use was devastatingly effective: only 3 of 800 bombers were lost, and Hamburg was pulverized.[2] German flak batteries were ineffective and their night fighters were useless—the guiding Würzberg radars were blinded.

During these early days chaff was more effective than it was subsequently. The Germans produced few types of radar system in World War II—they tended to stick with what had proved itself to work. The Würzbergs worked well, but they operated over a narrow bandwidth and this made them vulnerable to chaff. A 1942 British Commando raid to Bruneval, in Normandy, captured essential Würzberg components. These components revealed the narrow operating band and consequently, the following year, chaff rendered these radars useless. Here we have a typical example of EW: to defeat a particular radar system it is necessary to learn about its essential operating

---

[2]Pulverized but not destroyed. One of the enduring features of airborne bombing campaigns, in World War II and since, is the overestimation of their military effectiveness.

parameters, in this case frequency. The narrow operating band proved to be an Achilles' heel for this system. However the Germans learned how to improve their radars to deal with the problem (again, this is typical of the cut and thrust of EW evolutionary development). In desperation the Luftwaffe directed hundreds of engineers to find a way to defeat chaff (*Düppel*, in German) and announced a public competition with 700,000 Reichmarks prize money, tax free, for an effective solution. It has been argued that the concentration of effort on this one problem diverted German initiatives away from the direction of microwave radars, which the Allies correctly saw as the way of the future. Anyway, the Germans developed an add-on to their Würzbergs that enabled these radars to discriminate targets in Doppler—they could separate the slow-moving chaff clutter from the fast bombers by their difference in speed. Subsequent Allied bomber losses soared. Chaff was still effective against other radars, however, and the Allies distributed over 20,000 tons of it during the course of World War II.

Japanese World War II radars operated at long wavelengths and so the United States deployed long strands of aluminum chaff, up to 400 feet. This variety was known as *rope*, and was often supported at each end by cardboard kites, to keep it afloat for a longer period. It proved to be effective.

Nowadays chaff has become high tech. With so many radar systems out there, it is not advisable to simply assume that a certain type of radar with a known operating frequency will be illuminating your planes when they are within a specific airspace. Modern fighters and bombers are equipped with *radar warning receivers* (RWRs) that tell them that they have been illuminated by a radar transmitter beam, and at what frequency. If this radar is judged to belong to an incoming missile guidance system, then chaff will be dispensed automatically, without air-crew intervention. The chaff—nowadays metallized glass fibers rather than aluminum strips—is cut to the correct length on the fly (literally) as it is dispensed. However, the German solution still works: chaff is today effective only against dumb radars that do not apply Doppler processing.

In the present day, chaff is not the only defensive "expendable" that airplanes can disperse when under attack. Many active decoys have been produced. These are fired out of the plane when it is under missile attack. The decoys then transmit EM signals with, it is hoped, a signature that is close enough to that of the plane so that an approaching missile will switch targets. There is an IR equivalent that is perhaps better known through TV images: we have all seen footage of besieged airplanes shooting out flares to divert heat-seeking missiles. Such missiles zero in on the hot airplane ex-

Figure 4.1   A U.S. Navy Seahawk helicopter demonstrates countermeasure flares, designed to distract heat-seeking missiles. U.S. Navy photo by Photographer's Mate Airman Jonathan D. Chandler.

haust, and so a hot flare shroud is cast around the plane to draw the missiles away.

Active Jamming

As with chaff, active jamming is a form of EW originally intended to protect bomber formations, and today it is still associated mainly with air combat. Active jamming takes one of two forms: disruption and deception. A disruptive or *noise jammer* attempts to direct powerful electronic noise signals into an enemy radar, with the aim of blinding that radar. A deceptive or *repeater jammer* is sneakier: it tries to seduce the enemy radar.

In figure 4.2 you can see how noise jammers are deployed. Let's say that I am in a fighter plane and am looking for you, my enemy (just for a few pages!). You are piloting a bomber, which is also a *self-screening jammer* (SSJ). An SSJ does just like it says on the box: it sends out loud electronic noise with the intent of swamping enemy radar, so that the enemy cannot determine the bomber's location. From our earlier discussion on beam-

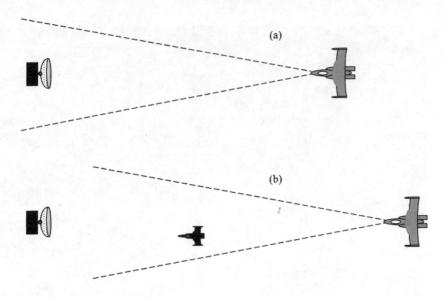

Figure 4.2 (a) An airborne self-screening jammer blasts noise into the antenna of enemy radar, so the radar cannot determine self-screening jammer range. (b) An airborne stand-off jammer blinds the radar so that it cannot determine the range or direction of a friendly plane.

forming you will appreciate that power density increases for narrower beams, and so to swamp a radar with noise, your jammer beam should be narrow. This poses a problem: how do you know where to point your jammer beam? Your bomber will have an RWR to tell you, and to tell your onboard integrated electronic defensive system, that my radar has illuminated you. The RWR determines the direction of my radar so that your jammer beam illuminates this direction only. There is a second reason why you would like your SSJ to have a narrow jamming beam: self-preservation. More on this topic shortly.

Jammer signals are one-way—they are not intended to return to sender. The range equation calculation of chapter 2 can be applied to jammers, and it is clear that one-way transmission yields a power density that decreases with range as R^{-2}—much more slowly than the R^{-4} behavior of a radar signal. This means that, if your SSJ has a transmitter with the same power as my radar transmitter, then it will produce a much stronger signal in my receiver than I get from your echo. I'm in a jam—swamped by your noise. Round 1 to you. In practice this range equation advantage does not mean that jamming always works, however, because you need to know my radar's operating band. Unlike in World War II, nowadays there are many, many radar systems out there operating in many different regions of the EM spectrum. So, you must transmit a wideband signal—much wider than most radar bandwidths, and this means that your power at any given frequency is diluted. For a given transmitter power, if you double the transmitted bandwidth then you halve the power at any particular frequency. It is like spreading butter: you can cover one slice of bread, or two slices with half the thickness. Even if you know my radar's operating band, I can concentrate transmitter power in a small part of that band, and so *burn through* your jam—see your echo despite the garbage that you are dumping into my receiver. If you *spot jam* in a very narrow bandwidth, to concentrate power where it is needed, I might be able to switch to a different frequency within my operating band to avoid you. If you try to follow me, I can switch frequencies quickly and randomly, so that your computers will not be able to find a pattern and predict my next choice of frequency. So it goes, with modern EW—a tail-to-tail chase over the EM landscape. Round 2 to me.

Another problem brings us to the interesting topic of your self-preservation. I hope that you are paying attention. Because you have transmitted a jammer signal, intentionally powerful so as to swamp my radar, you become vulnerable to *home-on-jam* missiles. These form one sect of the *beamrider* tribe of missiles that follow beams back to their source and zap the source. Such missiles do not require onboard navigation—you provide it for them. They head in the direction of increasing jammer power, and detonate when they hit the source.[3] Beamrider missiles make self-screening jamming dangerous, because your jammer range will exceed your own radar range. This means that you are making your presence known over a bigger volume of sky than you can cover with your radar, and so may be unwittingly alerting unknown enemies to your existence. (Hence, the desire to jam along a narrow beam, since this will limit the unknown volume that you illuminate.) Round 3 to me.

But you are not done yet. If SSJ is such a bad thing, nobody would do it. Beamriders are dumb (though, undoubtedly, they will be getting smarter as technology unfolds) and there is a way to confuse them. You have many friends in your bomber formation, and you help each other by coordinating your SSJ activities. You arrange to transmit your jammer noise while

---

[3]Think of a mosquito seeking a meal. She needs no fancy navigation or guidance schemes to find her prey—the algorithm she uses is simplicity itself. She heads toward higher concentrations of carbon dioxide, emitted by her prey on their breath. Home-on-jam missiles need be no smarter than this.

your neighbor is silent, and then he transmits noise while you are silent. This tactic is called *blinking*, and it totally confuses the traditional no-brainer beamriding missiles. They do not know which jammer source to home in on and end up passing harmlessly between the two of you. Round 4 to you.

In figure 4.2 you can see another type of noise jammer tactic: the so-called escort or *stand-off jammer*. Here a friend takes advantage of the longer range that usually applies for jammer signals, and sits behind the red zone—away from the scene of your attack. This renders him relatively safe from beamriders, while still providing you with jammer protection. Also, because you are not jamming, your position is not being divulged to me. I get jammed without learning anything about the bomber formation heading toward me, except perhaps that it is on its way. These days there are planes designed to do nothing other than act as stand-off jammers; their payload consists of massive transmitters that send out powerful noise signals over a wide area and over a wide bandwidth. Clearly, the frequency transmissions of these airborne ghetto blasters need to be coordinated with those of friendly aircraft that fly within the jammer beam, so that friendly radars can still work.

The action of *repeater jammers* is altogether more subtle. Once again you have been sent up in a bomber to attack me (can't we be friends?) and I am attempting to detect you with my radar. Your clever defense systems sense my radar pulses, and send back to me an amplified version of the same pulses. This requires some advanced processing capabilities—you need to discern the direction from which the pulses arrive, construct a similar waveform at the right frequencies, and send it back in an instant. The trick is that your repeater jammer slowly increases the delay time before returning the amplified pulses. My radar locks onto the stronger amplified pulses rather than the true echoes, and estimates range based on the delay time. By gradually changing the delay time, you are leading my radar away from the true range of your bomber. When far enough away, you stop transmitting, leaving my radar floundering around in empty space, looking for a target at ranges where there is none, like a bride who has been led to the alter and then jilted. This seduction technique is known as *range-gate pull-off*. It can be extended to give the impression that many bombers, not just one, have been illuminated by the radar.

There are equivalent repeater techniques for angle pull-off and velocity pull-off. It often helps to have a prior knowledge of the enemy radar system, and you will see that such knowledge was actively sought in World War II (I have already mentioned the Bruneval commando raid). Nowadays the information can be inferred in real time by the nature of the radar pulses that are intercepted by the jammer. For example, if a missile is heading toward your bomber and is tracking you with a conical scan beam,[4] then your deception jammer can infer the scan rate, based on the cyclically changing power of the tracker pulses. Your jammer then sends back pulses that vary in power, varying at the same rate as the conical scan. These fake echoes make the tracker think that it is off course, and so it adjusts to head toward the "true" target direction. Eventually, as with range-gate pull-off, the jammer dumps the missile radar in empty space. Seduced and abandoned.

Another technique is the so-called *cross-eye* jamming. Here your bomber has two jamming transmitters, one on each wing tip. The amplified echoes are now returned at very slightly different times. To my radar, the fake echoes appear to be from a different direction, because the wavefront (see note T1) formed by the combined transmissions from your wing tips has been bent, and so appears to originate elsewhere.

I hope that this short outline of jamming techniques has given you a whiff of what the Old Crows do for a living. As you might imagine, the field of EW is much more extensive than just radar jamming, but jamming is (and always has been) an important part of it. The opportunities for electronic deception open up more as technology develops, and the possibility of defeating such deceptions evolves in tandem. In World War II many of the modern deceptive tactics were a thing of the future—radar was still less than a decade old—but still the story of World War II airborne EW is an interesting one, which I summarize in the next two sections.

Defensive ECM and ECCM in England

At the beginning of World War II Britain and Germany knew of each other's interest in military remote sensing, and attempted to muddy each other's waters via espionage, false leaks, and other deceptions. A simple example: at the outbreak of war all German radio stations were switched to the same frequency, so that Allied planes could not use a radio frequency map to help them navigate over German air space. Here I will concentrate on the

---

[4]I mentioned conical scanning earlier in the context of sector scanning. The missile will head in a direction that corresponds to the axis of the cone, where the target is, which will maintain a steady echo power. Any cyclical variation in this power is then due to the missile being off course, and so adjustments are made to bring it back on track. These adjustments in reality can be, and need to be, made very rapidly if the aircraft target is throwing itself around the sky to avoid being hit.

airborne EW combat that took place over England during the Battle of Britain and during the Blitz of London.

First, in June 1940 the Germans built an extensive network of high-frequency navigation stations on the north coast of newly conquered France, in anticipation of air raids over England. Known as *Lorenz*, this system worked quite well, initially, during the Battle of Britain; any German bomber with a loop antenna could ride along the narrow Lorenz beams to their target. The British countered with the *Meacon* (*masking beacon*) system, once they had figured out what was going on. Meacon receivers picked up the Lorenz transmissions, relayed them by land line to Meacon transmitters, which broadcast the signals over a wide angular swath. The German bombers would receive both Lorenz and Meacon signals, and could not separate them, so that Lorenz navigation became useless. The extent to which bomber navigators were confused by this ruse can be judged from the fact that several Luftwaffe bombers became so lost that they landed on English airfields.

Next the Germans developed *Knickbein* (knock-knee), known to the British as *Headache*. This clever system consisted of two parallel beams broadcast from ground bases in northern France. One beam transmitted dots while the second transmitted dashes. The Knickbein beams would be directed to a target site, say London, and the Luftwaffe bombers would ride between the beams. When they were on target, their onboard Knickbein receivers emitted a steady tone; when they were off to the left they would receive dots, and when off to the right they would receive dashes. So the bombers were able to make course corrections depending on the tones they received. It is reckoned that this method directed aircraft with an azimuthal accuracy of 0.3°. Once the British tumbled to the Headache system, they responded effectively with *Aspirin*, which was basically the same as Meacon, but transmitted Headache pulses.[5]

In September 1940, at the height of the Battle of Britain, the Germans developed a devious navigation system that fooled the British for months. The *Ruffian* transmitters were high-frequency radio stations that sent propaganda broadcasts to Britain. But they had a second function that remained undetected for a while. Prior to a raid, the Ruffian transmitter beam would

---

[5]All this EW cloak-and-dagger stuff involves myriads of bizarre code names, and this continued throughout World War II to the present day. The main point to note here is that a clever technique would work for a while, producing tangible military benefits, until an effective countermeasure was developed by the other side.

be narrowed to 3°: this was enough for bombers to follow to a large target such as London, simply by tuning in to the broadcast. A cross-beam from another site in France would indicate when to drop bombs. Eventually, in England somebody saw through the deception—*Aha!*—and an effective counter to Ruffian was developed. *Bromide* transmitted Ruffian signals in the Meacon manner to disorient Luftwaffe bombers; additionally, fake cross-beam transmissions caused the bombers to drop their load over the English Channel. The Germans realized this deception by January 1941 and stopped using Ruffian.

The next German initiative consisted of loading all their various navigational aids in one squadron, KG-100, which utilized each method briefly and then switched. This must have been difficult to operate in practice, but this squadron of specialists made it work. They dropped incendiary bombs on the raid target (usually London), thus marking the site. Shortly afterward the main bomber formation would follow, see the fires caused by the incendiaries, and drop the main load of high-explosive bombs on the fires. Another *Aha!* in England, and the *Starfish* countermeasure was developed. The British started decoy fires in the open country south of London, on the likely route of the bombers toward their target. Seeing fires below, the bombers would dump their munitions harmlessly. An *Aha!* in Germany: the bombers learn to release their bombs on the *second* set of fires. An *Aha!* in London: the Starfish fires are moved to open country *north* of London.

Measure, countermeasure, and counter-countermeasure. While it worked, Starfish diverted about 50 percent of the Luftwaffe bombs from cities to open country. During the Battle of Britain, the Luftwaffe made a number of tactical errors, as we saw in chapter 1, and this blundering continued in the EW field. They would conduct experiments of new navigation systems within range of British electronic eavesdroppers, thus giving some indications of new developments. Also, they would set up navigation beacons in the afternoon, for a raid that night. RAF planes could fly along these beams and so guess the likely target.

Offensive ECM and ECCM against Germany

Eighteen months later, and the war in Europe was very different. Now the United States and Britain were bombing Germany, and the tables were turned. Now the Allies wanted to assist bombers in reaching enemy cities and in surviving enemy defenses, while Germany sought, with increasing desperation, the means to divert, deceive, and defeat air raids.

Initially, Anglo-American cooperation in EW activities began with the navies, perhaps surprisingly. Even before the Tizard technical mission, the U.S. Navy and the British Admiralty were putting their heads together to research EW techniques. The need for this became clear in 1942, when a clever move enabled two heavy German warships to survive a dangerous race up the English Channel to safety. The *Scharnhorst* and *Gneisenau* were bottled up in Brest harbor, on the northwest coast of France. The narrow Channel would have been a risky run for them, but for an EW coup. German transmitters sent noise into British radar receivers at dawn each day. The frequency, slowly fluctuating power, and timing of these jamming transmissions led the British to think that they were natural atmospheric interference, which just happened to be particularly strong at that time. Then on February 11, 1942 the *Scharnhorst* and *Gneisenau* slipped anchor and dashed up the Channel under cover of the jamming signals—completely undetected by British radar.

This and other developments convinced the United States to set up the Radio Research Lab, independently of radar research groups, to study ECM. This group grew to a peak of more than 800 personnel in 1944. One of the main concerns they had to deal with was the HS-293 German glide bombs. These weapons had sunk the Royal Navy ship HMS *Warspite* and the U.S. cruiser *Savannah*. Reports from the *Savannah*'s crew suggested that the glide bombs were directed onto their targets by an EM guidance system—this was confirmed when the British recovered one bomb in shallow waters off Libya, after a near-miss. Detailed U.S. analysis resulted in an effective jammer, so effective that the HS-293 was neutralized. Jammers onboard Allied ships during the D-Day landings ensured that none were lost to this potent weapon.

The main Allied effort, however, was in the air war over Germany. Research led the United States to produce training films, training courses, and aids to enable their radar operators to detect enemy jamming efforts, and invoke countermeasures to ensure that these efforts were overcome. Much more research was devoted to jamming German radars, however. In 1943 the United States deployed *Carpet I*, an airborne jammer system that filled up the German *Freya* radar receivers with noise. *Freya* radars were utilized to acquire (detect and locate) Allied bomber formations on their way into Germany. Once acquired, the detailed tracking of bombers, and the anti-aircraft defenses, were run by Würzbergs. Here *Carpet II* kicked in, sweeping noise power across the Würzberg operating band. This jammer system was so successful that the United States built 24,000 units for deployment onboard Allied planes. The Carpet jammers were first used on October 8, 1943 during a raid on Bremen, and the Eighth Air Force concluded that their losses had been reduced by 50 percent as a consequence. Later the Germans tried to de-jam their Würzbergs by frequency hopping. This countermeasure was countered by U.S. jammers that automatically followed each change of frequency.

The British developed the *H2S* (*Home Sweet Home*) magnetron-powered centimetric navigation system,[6] flown onboard Mosquito fighter-bombers, that successfully identified targets on the ground. H2S was deemed so successful that its use over Germany was forbidden, lest the Germans finally appreciate the importance of centimeter-wavelength devices. An RAF bomber shot down over Rotterdam, however, led to the Germans developing *Rotterdamgerät*, an imitation. However, by the war's end only a few had been produced. The wheel turns: a Luftwaffe Ju 88 bomber equipped with an early version of the *Lichtenstein* radar landed by mistake in England, baffled by the fog. Within a few weeks all Lichtenstein radars of this variety were jammed. Meanwhile British jammers were able to completely shut down Luftwaffe operations over England, by rendering their navigation devices useless.

The Allied ECM efforts were orchestrated, and symbolically this can be seen in the naming of two EW devices: *Oboe* and *Tuba*. Oboe was a navigational aid fitted onboard RAF Mosquitoes that directed them to targets up to 400 km distant, with an accuracy of about 110 m. The Germans learned how to jam Oboe (known to them as *Boomerang*) after a Mosquito crashed in January 1944 and was captured. The Allies responded by changing the Oboe operating band to centimeter wavelengths, powered by a U.S.-developed tunable magnetron that the Germans could not jam, while maintaining the old transmission wavelengths as a ruse. The U.S. Tuba system was a powerful (50 kW) ground-based noise jammer that blocked the signals of the airborne-interceptor version of the Lichtenstein radar. The Germans worked feverishly to develop a jammer that worked against centimetric radars, following the capture of a U.S. 3-cm radar. Eventually they succeeded in producing a few systems, but too little, too late.

---

[6]Several physicists who would later become well known within their field, such as the radio astronomer Bernard Lovell, worked on the H2S system. In 1942 the pupils at Malvern College, an exclusive private school, were booted out of the school buildings and the newly established Telecommunications Research Establishment, which developed H2S, moved in. The British cosmologist Fred Hoyle and the American theorist Julian Schwinger are two of the many physicists who worked on radar during World War II.

After the war the Allies interviewed German engineers, and analyzed the Reichforschungsrat (German scientific R&D office) records. They wanted to know how effective the Allied ECM programs had been in thwarting German air defense systems. They concluded that countermeasures had reduced the effectiveness of the entire German radar network by four-fifths. German AAA batteries had been effectively blinded and could not detect Allied bombers by radar. To deny the Allies this knowledge, the batteries were ordered to fire anyway, though this was otherwise a waste of ammunition. Then a frantic *volte-face*: the batteries were ordered *not* to fire unless they had positive visual contact. Such was the importance of radar, and the effectiveness of Allied countermeasures, that more than 4,000 German engineers had worked on methods to de-jam their radars. This is more than five times the peak number of U.S. engineers employed on countermeasure development.

In the 60 years since the end of World War II, radar and ECM development have not stood still, of course. They have moved forward in tandem, locked in combat. A major escalation in Cold War EW occurred in the skies above Vietnam, where Russian-built surface-to-air missiles attempted to shoot down U.S. Air Force bombers, which attempted to avoid that fate. This was the era when RWRs were introduced, and also defensive jammers, such as the SSJs discussed earlier. The *Wild Weasel* antiradiation missile (ARM—the type of missile that I have hitherto called a "beamrider") came in, and spawned a long line of descendants.

The jewel in the crown of EW effectiveness came over Iraq during the first Gulf War, when SEAD (Suppression of Enemy Air Defenses—not just weasels but a whole menagerie of radar jammers and other ECM, plus ECCM and EM- and IR-guided smart weapons) rendered Iraqi air space a no-fly zone for Iraqi planes, and rendered Iraqi radars blind to Coalition air and ground force movements. Several of the advances that I will discuss in a later chapter were first deployed in the Gulf War.

Anomalous Propagation

To quote from Monty Python: "And now for something completely different." Anomalous propagation is not some bizarre, mind-boggling new way of reproducing the species; instead it refers to the seemingly mundane fact that light and sound waves do not necessarily travel in straight lines. This physical phenomenon originates because the medium, be it air or water, through which the waves travel is not uniform. The consequences for military tactics are significant, which is why I am including this phenomenon.

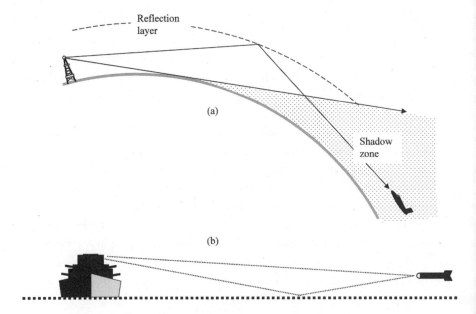

Figure 4.3 (a) Radio waves reflect off the ionospheric boundary layer (ray shown, with arrow) permitting over-the-horizon detection. Without this layer, or for radar wavelengths that do not reflect at this boundary, the earth's curvature creates a shadow zone (shaded). (b) Multipath reflection. Echo pulses arrive at the radar (left) via a direct path and also via a reflected path, in this case from the sea surface.

A technical appreciation of anomalous propagation will be bestowed on anyone who chooses to read note T21.

First, I will provide two important examples of how reflection of EM rays influences military operations. I will make an approximation here: microwaves travel through air in straight lines (nearly true[7]). So, I am ignoring atmospheric refraction. This assumption simplifies my explanations, and this is the only reason I make it: including refractive effects would not change the arguments. The two examples that interest us are *ionospheric reflection* and *multipath*. These phenomena are illustrated in figure 4.3.

The earth's curvature limits line-of-sight radar range. Suppose that a radar antenna is near sea level, at elevation 10 m, and is looking for an airplane

---

[7]There is downward curvature due to the change in atmospheric density with altitude, as discussed in note T21. It turns out that, for calculating the radar horizon, we can take this curvature into account by pretending that the earth's radius is 4/3 larger, and by treating the rays as straight lines.

target at altitude 100 m over the sea. Because of curvature, the airplane will appear to be on the horizon when it is separated from the radar by 53 km. At longer ranges the target is below the horizon. This shadow zone is illustrated in figure 4.3. If our radar is at greater altitude, say 100 m above sea level, and the target plane is flying at 3,000 m altitude, then the shadow zone begins at 260 km range. So how could the ground-based U.S. navigation radar Loran, described in chapter 1, cover all of central Europe, which extends for thousands of kilometers? The answer is also shown in figure 4.3. Loran and many other long-range radars operate at long wavelengths (radio frequencies). Such wavelengths lead to poor resolution, but, on the other hand, they have very little atmospheric attenuation, as we saw earlier. In addition, they reflect off the so-called E layer of the ionoshere, located about 150 km above the earth's surface. At these altitudes many of the atoms that constitute the atmosphere are charged ions, and form a discrete layer that effectively reflects radio waves (shorter EM waves, such as microwaves and visible light, pass through the ionosphere as if it wasn't there). Because of this reflection, radio waves can be bounced off the E layer and onto an OTH (over-the-horizon) target, as shown. In fact, several bounces can be achieved, so that long-wave radar coverage can extend a significant way around the globe, as every radio ham knows. The E layer is surprisingly stable, but there is one big problem. Underneath the E layer is the D layer, which strongly absorbs radio waves. This D layer depends on sunlight for its existence, and disappears at night. (So long-range AM radio stations have to boost their transmitter power at sunrise to counter the absorption of radio waves by the D layer, and ease off the power at sunset.) Because of the D layer, Loran worked only at night.

Long-range sea-skimming missiles take advantage of the earth's curvature shadow zone by flying in low. This keeps them below the defensive search radars onboard naval vessels. You will no doubt have noted already that many warships have tall masts with radar antennas placed high up—clearly this increases the radar range by increasing the radar horizon.

There is another reason why sea-skimming missiles are favored, which is connected with reflections of radar waves off the sea surface. As illustrated in figure 4.3, the rays that are transmitted to a sea-skimming target can take two paths: either direct or reflected off the sea. If the sea is calm, this reflection can be strong—almost as strong as the direct ray. Similarly for the echo. Because of the reflecting surface, the radar receives two echoes from one target, and so this phenomenon is called *multipath*. Since the missile target is at low altitude, there is very little difference in the path lengths between the direct and reflected rays. This is bad news for the radar, because it makes detection of the missile much more difficult.

To understand why multipath makes the detection of sea-skimming missiles difficult, we need to recall two topics discussed earlier: coherent integration and wave interference. The two echo waves returning into the ship radar receiver antennas add together—this is coherent integration (provided by nature—in this case the radar signal processor does not perform the integration and, indeed, does not want it). The point is that integration does not necessarily increase the echo power, because of wave interference, as we saw in figure T1.3. The combined wave is alternately stronger and weaker than the direct echo alone, so the target fades in and out as it approaches, making tracking difficult. Also, the direct and echo waves arrive from slightly different directions, and this means that the combined wavefront of the two echoes is bent. To the radar receiver they appear to originate from a target that is higher up than the real missile, then lower down—below sea level sometimes—then higher up again, and so on. As the missile approaches, the poor tracker radar that is trying to follow it starts nodding, because (due to wave interference of the direct and reflected echoes) it thinks that the missile is undulating up and down. This nodding behavior is a well-known problem of ship-borne radar trackers that are trying to lock onto incoming sea-skimming missiles.[8] Sometimes the flickering effects of multipath echoes are enough to break lock, and the tracker loses the missile.

Multipath is a particularly useful smoke screen for incoming missiles because it is predictable: if the sea is calm then multipath reflections will occur. Other meteorological conditions give rise to anomalous microwave propagation and strange echoes (for example, there are microwave mirages, just as there are optical mirages in the right weather conditions), but most of these are not predictable and so cannot be adapted for tactical use. For acoustical waves propagating through seawater, however, there are many phenomena, not just multipath (known as *Lloyd's mirror* to physicists and sonar engineers alike), that are predictable and exploitable.

---

[8]Twenty years ago I read a Russian textbook that gave a full and rigorous mathematical analysis of this nodding behavior. The Russians have always been very good at mathematically analyzing engineering problems. You might expect that the tracking radar would estimate the missile altitude to be somewhere between its real altitude and that of its reflection (if the missile is 5 m above sea level, then its reflection appears to be 5 m below sea level). Perhaps surprisingly, however, due to wavefront distortion the tracker can estimate the missile to be at altitudes well above and below this extent.

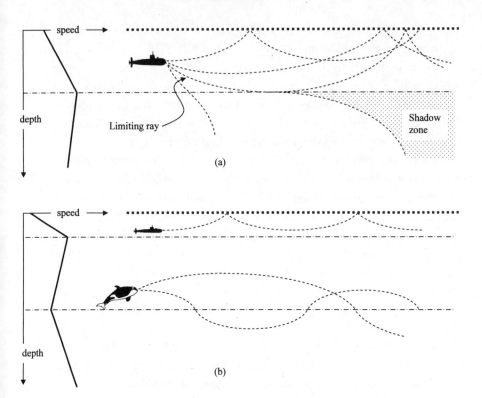

Figure 4.4 (Left) *Speed of sound versus ocean depth profiles. (a) For the profile shown there is a shadow zone, which cannot be penetrated by the submarine's sonar. (b) For this profile there is a deep sound channel: the whale can use this channel to communicate over long distances.*

As we descend beneath the ocean surface, water temperature, salinity, and density all change, and these changes cause the speed of sound in sea-water to become depth dependent, as discussed in note T21. Depth dependence varies geographically, and in any given location it may change slowly in time. A typical profile of speed versus depth is shown in figure 4.4. In this case, the speed of sound increases initially, as we descend, to reach a maximum at a few hundred meters depth. Then the speed falls away to a minimum at a depth of perhaps 1,000 meters, before rising again. From note T21 we saw that sound rays bend toward regions of lower sound speed, and so from the speed profile of figure 4.4 you can understand why the acoustical rays bend as shown. For a submarine operating near the ocean surface, a shadow zone is imposed by the speed profile. Sonar pulses from the sub cannot reach this region, which is delimited from above by the depth of maximum sound speed, and from below by a *limiting ray* of sonar acoustical power that just grazes this layer. If you ponder figure 4.4, and keep in mind that sound travels toward lower speeds, then you can understand how a shadow zone arises.[9]

If we descend a little deeper, so that we are below the depth of maximum sound speed, we encounter (in fig. 4.4) a substantial cetacean—in this case, an orca whale. Echolocating whales know all about *deep sound channels*, and our orca is in one. No shadow zone here: sound waves emitted from a source at these depths oscillate around the depth of minimum speed, as shown. Because of this behavior, the acoustical power does not dissipate as fast as normal. Recall from our range equation calculation that power density usually falls away as the inverse square of distance, R^{-2}, as the waves spread out through a volume. (The two-way behavior goes like R^{-4}.) Here the spreading is only in two dimensions, since the rays are confined to a narrow channel in depth. This means that the power density falls off like R^{-1} as the sound travels away from its source (and so like R^{-2} for two-way travel). So, because the acoustical power decreases much more slowly with range, the maximum effective range of a sonar at these depths is much greater. This makes deep sound channels dangerous places for submarines, but our orca makes use of them for long-distance communication. Deep sound channels are the telephone booths of the cetacean world, especially for low-frequency one-way communication[10] (imagine an orca mom calling her kids home from school).

In figure 4.4 you can see that there is a surface duct: the speed profile leads to the channeling of sound trapped in this layer. This region is not as efficient as deep sound channels for transmitting acoustical power over long distances, but it is better than normal (i.e., better than straight-line propagation). It is less efficient because acoustical power can leak out of the duct from below (for rays that are beneath the limiting ray—see fig. 4.4) and because reflection off the ocean surface is not perfect. So you see that, depending on the local speed profile, the range of a sonar system depends very much on how deep it is, and how deep the target is.

Different speed profiles lead to different results, of course. For example, if the speed decreases continuously with depth, then there is a shadow zone

---

[9]I find it helpful to consider the speed profile (fig. 4.4, on the left) as a hill, along which we roll golf balls. The golf ball trajectories will look rather like those of the acoustical rays.

[10]Low-frequency sound waves are attenuated much less strongly in water, you may recall, than are high-frequency waves.

near the surface, no matter how deep the sonar transmitter may be. (I'm going to let you show that this is the case. You now possess the know-how.)

Submarines on Top

Antisubmarine warfare (ASW) has a long history, and reached its peak intensity—at least in terms of R&D, of cost, and of priority—during the Cold War. If you were to talk to a sailor or airman whose job it is to conduct ASW (whether onboard a surface warship that is hunting subs, or a helicopter pilot who drops sonobuoys, or a sonar operator who stares at the complex sonar echoes displayed on computer screens), he will likely tell you that modern subs are practically invisible and unstoppable. It is a wonder that subs are ever detected, let alone destroyed. If you were to talk to a submariner, he would likely tell you that subs are vulnerable beasts, so great are the variety and number of remote sensing devices dedicated to their detection, and so potent are the weapons available for their destruction. It is a wonder that submarines can function at all. The truth, in the modern era, is between these extremes, though historically subs have generally had the upper hand. Even knowing this, when talking to ASW servicemen or submariners you may find yourself drawn to either extreme view, because both groups can marshal strong arguments to support their positions.

First, look at the problems faced by ASW personnel. Except for one brief period in World War II, submarines have been a potent threat in warfare, and have determined strategic and national policies at the highest level. The significance that is attached to the strategic capabilities of a submarine fleet seems to be out of all proportion to the number of submarines that a nation possesses. For example, during the convoy war in the Atlantic during World War II the number of German U-boats (*Unterseeboot*—undersea boat—sometimes German is delightfully literal) never exceeded 240, yet the Royal Navy resources that were allocated to combat these submarines included 875 ASDIC-equipped escort ships and 300 Coastal Command aircraft. Additionally, there was substantial U.S. and Canadian air and marine assistance, plus an enormous number of R&D engineers and military analysts on both sides of the Atlantic who spent the war combating U-boats. Their efforts were rewarded by success, as we will see, though this was brief: submarines quickly regained the advantage even before the end of the war. This advantage was maintained for decades afterward, in an era—the Cold War—where submarines became the principal offensive platform for both the United States and the Soviet Union.

Given the small numbers of subs, we can appreciate the difficulties that face ASW warriors. The oceans cover 70 percent of the earth's surface, and all but the shallowest seas are open to submarines. The oceans comprise approximately one billion cubic kilometers of saltwater, and submarines can lurk in all but the deepest.[11] The most common sensors utilized in submarine detection—sonar systems—have a range that is short compared with radars operating in air: a few kilometers (for active operation) or tens of kilometers (passive). Given the vastness of the available oceans, the submarine's autonomous and self-reliant nature, and the short range of ASW sensors, it is a daunting task indeed to detect (let alone destroy) an enemy sub.

Until the 1950s submarines were utilized mostly for coastal and fleet protection, and for destroying enemy shipping. The Kaiser's use of unrestricted submarine warfare in World War I against Allied shipping led directly to U.S. participation; the continuation of German unrestricted submarine warfare led to a delay in concluding that bloody conflict. Again in World War II the Germans turned to their U-boat fleet to bring Britain to heal. "Wolf packs" of U-boats caused mayhem in the Atlantic, sinking tens of thousands of tons of merchant vessels and warships each month. The problem grew acute: sinkings increased until May 1943 and Britain's lifeblood—supplies from North America—was hemorrhaging fatally. Only a massive Allied effort was able to reverse this trend. We have seen that the introduction of centimetric radars permitted detection of U-boat snorkels. There were many other innovations that contributed significantly:

- Depth charges were the main weapon utilized in destroying U-boats, and these were improved.
- The United States invented sonobuoys,[12] and these permitted a greatly enlarged area of ocean to be covered by sonar detectors.
- Leigh lights were very powerful searchlights, slung underneath ASW aircraft. These lights were effective once the air crews learned to switch

---

[11]Submariners become more than a little apprehensive if they are at depths below a few hundred meters.

[12]*Sonobuoys* are hydrophones (underwater microphones) dropped into the water from planes or helicopters. They listen for a submarine and relay their echo signals by radio link back to the aircraft. Sonobuoys are usually passive, but some of the larger and more expensive ones are active—they transmit acoustic pulses and monitor the echoes. There are sophisticated *dipping* sonobuoys (highly directional, usually active—and expensive) that are suspended from a helicopter and later retrieved, but most sonobuoys are cheap and expendable—they float for an hour or so and then sink to the ocean floor. Some parts of the seabed must be littered with tens of thousands of these listening devices.

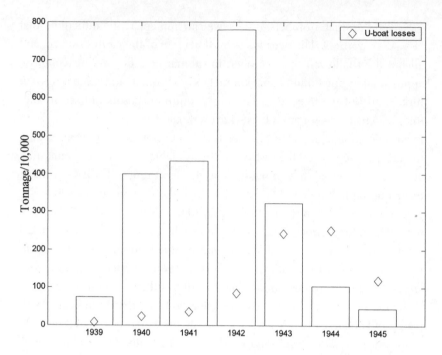

Figure 4.5 *U-boat losses for each year of World War II (diamonds). The bar chart shows the loss of Allied merchant shipping—mostly British—in units of 10,000 tons. So, in 1942 almost eight million tons of merchant shipping was sent to the bottom. Over half of these losses were caused by U-boats.*

them on only during the final approach to a U-boat. Leigh lights denied U-boats the safe cover of night for recharging their batteries, for which they needed to surface.

- The Royal Navy introduced *Huff-Duff* (High-Frequency Direction Finder)—an accurate and effective direction finder that homed in on U-boat radio signals.

- Military analysts applied mathematics to help sharpen the Allies' response to the U-boat threat by analyzing the results of various tactics.[13]

---

[13]One simple example of the application of math to the U-boat war involves convoy size. If, say, five escort ships are protecting a convoy of 40 merchant vessels from U-boat attacks, they form a ring around the convoy, with a certain distance between the escorts. Now increase the numbers by a factor of nine: 45 escorts form a ring around a convoy of 360 merchantmen. The area occupied by the convoy has increased by a factor of nine, but the perimeter is only three times as long. So the escort warships are three times closer together—making it less likely that a U-boat can penetrate. The Royal Navy was very slow to appreciate the benefits of large convoys.

Figure 4.6 *"U-118 under attack by aircraft from* USS Bogue (CVE-9), *June 12, 1943. LTJG Fryatt's depth bombs straddle U-118. Splashes from his turret guns can be seen as the Avenger pulls away after the attack. Two crewmen can be seen seeking shelter behind the conning tower. U-118 is trailing oil after previous attacks by LTJG Stearns and LTJG Fowler." This U-boat was sunk, just off the Azores, at a time when the Allies were turning the tables in the U-boat war. I am grateful to Captain Jerry Mason (USN, retired) for providing this dramatic photograph.*

Results were dramatic: the huge effort resulted in a drastic increase in U-boat losses after May 1943, with a consequent reduction in convoy damage. (The dry statistics are conveyed in fig. 4.5. The more evocative photograph of fig. 4.6 provides a better indication of what was at stake.) Note, however, that, over the course of the Atlantic war, four Allied merchant ships were sunk for every U-boat that was sunk.

This enormous and seemingly disproportionate ASW effort was merited by the importance of keeping shipping lines open across the North Atlantic. The victory of ASW was short lived, however. Even before the end of the war, new German U-boats were introduced that were able to beat the ASW apparatus. These so-called Type XXI U-boats were faster and quieter.

Happily they were also too few and too late—the Third Reich was on its last legs by the time Type XXIs became available. But the writing was on the wall: subs were back.

Two developments brought submarines to the forefront of Cold War strategy: nuclear power and nuclear ballistic missiles. A submarine that was powered by nuclear energy did not need to surface to recharge batteries—it could remain underwater for many months—and its range of operation was limitless. As they grew larger (doubling to about 3,000 tons by the 1960s) submarines housed ballistic missiles (SLBMs, or submarine-launched ballistic missiles) that could be launched from the ocean depths. The subs were free from the constraint of operating on the surface and near enemy vessels or shores: an SLBM targeting New York could be launched at depth from the mid-Atlantic. How on earth could this threat be countered?

Submarines grew larger still (up to 30,000 tons by the 1980s, for some of the Soviet behemoths) and more dangerous. The Alfa-class Soviet subs had titanium hulls (and were very fast, at 45 knots). Apart from being hideously expensive, such machines could go to depths where no Western sub could follow. The fact that Soviet submarines were accident prone, and cost many Russian mariners their lives,[14] was of little comfort to the West. How could their threat be overcome?

ASW

Yet, as a result of huge and continuing effort, it was overcome—or at least curtailed. Now we come to the submariners' version of events: that subs should barely be able to function given the massive resources arrayed against them. Because of the Cold War threat posed by submarines, the West, in

---

[14]Many readers will now know, thanks to a Hollywood film, of the K-19 accident in 1961. This Soviet sub was nuclear powered and housed ballistic missiles. The nuclear reactor coolant system broke, threatening nuclear meltdown, and some of the crew fixed it enough to get the K-19 to port. All crew members suffered overdoses of radiation—massively so, for the eight who effected the repairs, and who died within days. In 1969 the K-19 accidentally collided with a U.S. sub in the Barents Sea. Many Soviet nuclear subs suffered reactor failures. (The problems persisted into the post–Cold War era, with all 118 hands lost when in 2000 the *Kursk* Oscar II class submarine went down following a chemical explosion equivalent to 100 kg of TNT.) The United States lost more than 200 servicemen in two nuclear sub accidents, but Soviet losses from such accidents are certainly much higher. Bizarrely, given their importance, the Soviet fleet of nuclear submarines was manned by recruits rather than by dedicated professionals (except for the officers). On one apocryphal occasion a British ASW aircraft was photographing a surfaced Soviet vessel—a common Cold War encounter—and found large lettering in white paint daubed on the side of the hull. Back at base the Cyrillic script was translated: "We want to go home."

general, and the United States, in particular, has invested trillions of dollars to find ways of detecting and destroying Soviet submarines. Many thousands of servicemen and R&D engineers have spent their entire careers hunting submarines. Nowadays, sonar has progressed way beyond the simple ASDIC systems of World War II, as you soon will see.

Radar and sonar aside, ways to detect submarines do exist: MAD and satellites. MAD stands for magnetic anomaly detection. The idea is old and simple: the local magnetic field at the ocean surface is disturbed when a large metallic object passes by. So, surface ships can detect nearby submarines by monitoring magnetic field changes. Satellites can detect submarines from space. I do not mean that they can photograph a sub that is on the surface—no doubt they do—but rather that satellites can detect *submerged* subs from space. They do so by applying careful signal processing to the echoes that onboard imaging radar obtains of the ocean surface. A submarine moving underneath that surface produces internal waves that percolate upward, and the surface pattern of these waves is characteristic of a submarine source—surface vessels produce a different-looking wake. So, the eye in the sky sees the motion in the ocean. With enough processing power, a sub-hunting satellite can cover an awful lot of aquatic territory in this manner—assuming that the seas below are not too rough.

The main effort in detecting Cold War submarines, though, was invested in sonar. Ships could tow long strings of transmitters that effectively produced a high-powered and highly directional source of acoustic radiation. They could suspend a similar linear array vertically. Usually, these arrays were passive "eyes" rather than active "ears." As we have seen, range is longer for passive detection (one-way propagation), and of course the sonar does not give away its presence to the submarine. There are methods of processing the echo data from a moving sub—data that may be gathered over hours rather than over seconds, as with radar—that enable a distributed array of receivers to estimate accurately the target range, direction, depth, speed, and classification. Nuclear subs make a noise that is different from that made by diesel-electric subs,[15] small subs sound different from big ones, different classes of sub make different engine noises, and so forth. A menagerie of sonobuoys, of bewildering variety, has been developed to work

---

[15]The sound produced by a nuclear-powered submarine cannot be switched off, since the reactor cannot be switched off, whereas diesel-electric engines can be shut down completely. Hence a small hunter-killer diesel-electric sub is harder to detect, when running on batteries, and impossible to detect if the crew are prepared to remain stationary and silent for many hours.

alone or in unison. Dropped randomly into the water at a likely spot, suspended at different depths from floats, they form a three-dimensional array of sensors that is highly sensitive and can be deployed in minutes. If a submarine is not detected, the ASW aircraft can repeat the deployment with more sonobuoys somewhere else.[16]

No survey of Cold War sonar would be complete without SOSUS. The Sound Surveillance System was a network of seabed passive acoustical receivers, very sensitive, placed at strategically important points around the world. Begun in the early 1950s, this system expanded right until the end of the Cold War in the 1980s. By that time, SOSUS arrays were strung along both the Atlantic and Pacific coasts of the contiguous United States, in the Mediterranean, and from the Aleutian Islands off Alaska to Japan. There were further arrays in areas of heavy submarine traffic: strung along the ocean bed between northern Norway and Svalbard (not far from the White Sea and the major Soviet naval base at Murmansk, east of Finland) and strung along the GIUK gap (Greenland–Iceland–United Kingdom).

Most of the oceans in the Northern Hemisphere were patrolled by ASW aircraft from all the major Western countries. It has been claimed that, by the end of the Cold War, the location of all Soviet nuclear subs was known, all the time: 24/7 every week of the year, continuously for many years. Once a submarine had been detected, classified, and identified, its unique acoustical signature could be recorded, and it would be tracked by SOSUS, and by ASW ship and aircraft patrols. Presumably these data were augmented by satellite sensors and the whole panoply of nonacoustic ASW detection apparatus.

Fortunately the Cold War remained cold, and it was not necessary to destroy enemy submarines. Had this been required, it might have been achieved by firing depth charges, much improved since World War II but basically the same type of munition. Or the subs may have been destroyed

---

[16]I recall an incident that occurred when I worked in the sonar signal processing field, toward the end of the Cold War. The Royal Navy was patting itself on the back for having recovered a Soviet sonobuoy. This was no easy achievement, because most sonobuoys are small and they remain afloat for only an hour or so at the most. The captured sonobuoy was pulled apart and it was found that, though this was one of the cheapies that were dispensed by the tens of thousands, it was capable of detecting very-low-frequency sounds. Such a capability is difficult to achieve, and without it the sonobuoy could have been manufactured for a small fraction of the cost. This finding emphasized the feeling that was around at the time that (diesel-electric) subs running on batteries were now so quiet that they gave themselves away only at the lowest end of the frequency spectrum.

Figure 4.7 *Cold War adversaries. (a) Los Angeles–class fast attack nuclear submarine USS Asheville. U.S. Navy photo by Journalist 2nd Class Zack Baddorf. (b) Soviet Delta-class ballistic missile submarine. The first Delta I appeared in 1972. SOSUS first detected a Delta at sea in 1974. U.S. Navy photo.*

by smart mines, lurking on the seabed until they detect a sub passing overhead, at which point they would "rise and surprise." Or by one of the many new homing torpedoes—large, fast, intelligent, and expensive.

As for the tactics that a submarine captain could call on to evade detection, we have already examined three. He might be able to hide in shallow water, which is noisy and emits a lot of confusing reverberation echoes—or he may hide opportunistically among some other source of strong reverberation, such as sea ice. More commonly he may be able to hide in shadow zones that arise because of anomalous propagation of acoustic radiation. If he has a periscope up, he will drift with the speed of surface waves, so that his periscope does not stand out in the signal processing of a Doppler radar. In addition to these tactics, the captain can always rely on one advantage that submarines possess in their battle with ASW forces: all modern submarines are now acoustically stealthy—the result of decades of evolutionary "quieting." Diesel-electric subs are all but silent when running on batteries. When detected by enemy sonar, the sub captain can shut down all engines, fans, refrigerators, and other onboard noise sources. The sub must be stationary—there can be no hull noise from sliding through the water. The crew must be silent; there must be no conversation, and they can make only careful movement on silent shoes—ASW ears nowadays can detect faint "transient" sounds such as a door shutting. To a sonar sensor, this quieting must seem like a Klingon cloaking device switching on.

BIBLIOGRAPHY

The Association of Old Crows maintains a Web site where you can find more information about their aims and activities: www.myaoc.org/eweb/StartPage.aspx.

Barton, David. *Radar System Analysis*. Artech House, Boston, 1976.

Cote, Owen R., Jr. *The Third Battle: Innovation in the Navy's Silent Cold War Struggle with Soviet Submarines*. Newport Paper No. 16. U.S. Naval War College Press, Newport, RI, 2003. This interesting and comprehensive account of ASW during the Cold War is also online at: http://handle.dtic.mil/100.2/ADA421957.

Harris, D. B., Lorenzen, H. O., and Stiber, S. History of electronic countermeasures. In *Electronic Countermeasures*. Peninsula Publishing, Los Altos, CA, 1978. Very strong on World War II ECM and ECCM.

Hepcke, Gerhard. *The Radar War*. The text is available online at: linepc.ath.cx/topsecret/radarwar3.pdf.

Miller, J. *Lockheed Martin's Skunk Works*. Voyageur Press, Stillwater, MN, 1996.

Palmer, Alan. *Victory 1918*. Grove Press, New York, NY, 1998.

Price, Alfred. *Aircraft Versus Submarines: The Evolution of the Anti-Submarine Aircraft 1912 to 1972*. Naval Institute Press, Annapolis, MD, 1973.

Taylor, Frederick. *Dresden*. HarperCollins, NY, 2004.

Several very extensive Web sites are dedicated to U-boats; for example, www.uboatarchive.net; www.uboat.net.

5 . . . .

**Mapping:
Hearing
the Picture**

Most modern radar screens present the data on a display laid out in the form of a two-dimensional map. Perhaps the display shows range against azimuth angle, as in figure 5.1—this is typical of surveillance radars that are searching the sea surface, or looking for incoming low-level missiles or airplanes. Some displays show plots of range versus speed, or azimuth versus elevation. The purpose of these types of displays is to convey the target information to the radar operators in the most efficient manner. Thus, in a military context, the radar operator will want to know the location of a hostile target relative to other hostiles and relative to friendly forces. So the target is projected onto a two-dimensional map that allows the operator to quickly assess the situation. For air traffic control (ATC) radars the targets are not hostile, they are cooperative and they want to be seen by the airport radar. The same type of display is used, however: the targets are plotted on a map of the airport and surroundings, so that the ATC operators can guide aircraft safely down onto the correct runway. Both ATC and military radars these days convey further information, by annotating the maps and by color-coding the displayed targets, so that the operators know about target altitude and speed, and possibly about target type.

In all these cases, the background map is just that—background. It places the targets in context, and is not of itself very interesting to the radar operators. So, such background maps do not have to be very detailed. Sometimes

Figure 5.1 *Typical surveillance radar display. Here we see targets on the screen, displayed on a plot of range versus azimuth angle. U.S. Navy photo by Photographer's Mate 3rd Class Andrew S. Geraci.*

the background is simply a contour or street map—this works if the radar is in a fixed location, as for ATC radars. In other cases the map is simply the set of clutter echoes for each resolution cell. That is, the echoes from each cell (deemed to be clutter, and not targets) are displayed in their proper location, but dimmed or plotted in a different color that allows the targets to stand out clearly.

Mapping radars stand apart from the radars described so far in this book, and the displays of these radars consist of far more than simply the background clutter maps of conventional radar displays. One man's clutter is another man's target. For imaging radars, the topographical features *are* the targets, and these are resolved and displayed with much, much higher resolution than is conventionally possible. I will explain how these radars work.

Enter SAR
SAR stands for Synthetic Aperture Radar, for reasons that will soon be abundantly clear. SAR systems, and their acoustic counterparts *sidescan sonars*,

represent the way that humans do remote sensing.[1] When it is time for me to stack up our human remote sensing achievements against those of the world champions—the bats—it will be SAR systems that I send forward to represent humanity.

Human remote sensing began rather later than most radar development, in that most people consider it to be a post–World War II innovation. It is unusual also in that its origin is, by general acknowledgment, the unique achievement of one country. As we saw in chapter 1, most of the radar techniques in use today had their origins in World War II and most were invented several times over in different countries. SAR came into being in the 1950s, in the United States. Carl Wiley of the Goodyear Aircraft Corporation first thought of it in 1951. He saw a way to use the Doppler effect to improve radar cross-range resolution.[2] At about the same time, researchers at the University of Illinois hit upon a different way of processing Doppler information to improve resolution. In the next few paragraphs I will show you how SAR works, without mentioning Doppler very much, since the explanation in terms of Doppler is rather mathematical. You should bear in mind, though, that the ultimate reason why SAR achieves high angular/cross-range resolution is because it exploits the Doppler information contained in the radar echoes. A great deal of processing is necessary to turn the echoes into a SAR map, and it was not until 1957 that the first fully focused (i.e., processed) SAR map was produced. Development in SAR has continued since 1957; today SAR systems are still being developed, and indeed they are now a main strand in the fabric of radar development. Improvements in electronic components and in computer power over the decades have

---

[1]Not quite true, since a lot of satellite mapping is done at optical and IR wavelengths. Most people are rightly impressed by the very-high-resolution images of the earth obtained by cameras onboard earth-orbiting satellites and these, plus IR images, constitute an important class of remote mapping sensors. The resolutions obtained by optical and IR cameras are much higher than can be obtained by radars, because these EM waves have much shorter wavelengths than microwaves do (recall note T4). So, you might well ask, why bother with SAR at all? We saw the answer in chapter 4: optical wavelengths are scattered and attenuated by atmospheric aerosols and water vapor. Clouds will obscure the view of an optical spy-in-the-sky. Also, optical cameras do not work very well at night. Microwaves can cut through the clouds and enable us to see in the dark.

[2]*Cross-range resolution* is another way of saying *azimuth angle resolution*. If a radar antenna has an azimuth resolution of $\delta\theta$ then it has a cross-range resolution of $R\delta\theta$ at range R. That is to say, two targets at the same range, separated by a distance of at least $R\delta\theta$, will be resolved. So, for example, if $\delta\theta = 1°$, then the cross-range resolution at a 10 km range is 175 m; at a 100 km range it is 1,750 m.

proved to be hugely important in advancing SAR. Future hardware improvements, in tandem with theoretical work and new processing algorithms, will take SAR and its new child STAP—discussed in chapter 6—to the forefront of our remote sensing capabilities.

Note that SAR improves cross-range resolution; it does not improve range resolution at all. However, the main limitation in conventional radar mapping without SAR processing is that the resolution cells at long range are very large in the cross-range dimension. We have already seen how range resolution has been greatly improved by the technique of pulse compression (note T20): this technique can give us resolution cells that extend only 1 meter in range, or less. Some experimental systems can achieve better range resolutions than this—in the recent radar literature I have seen reports of a remote sensing radar with a range resolution of 10 cm (about 4 inches). Compare these numbers obtainable for range resolution with the numbers that we obtain for cross-range resolution without SAR. The beamwidth of a radar depends on transmitted wavelength λ and antenna width D, as we have seen. The angular resolution is just the beamwidth, so $\delta\theta = \lambda/D$. So, the cross-range resolution (which I will denote by δR_x) at range R is $\delta R_x = R\delta\theta = R\lambda/D$. Now for an airborne radar system we cannot have very large antennas, and for long-range operation we must use microwaves with wavelengths of at least a couple of centimeters (smaller wavelengths are attenuated too strongly), and so there is a practical limit to the beamwidths that can be achieved. So, beamwidths and angular resolutions of $1°$ or so are about as good as we can expect for an airborne system, and perhaps ten times better than this (one-tenth of a degree) for ground-based, stationary radar systems. But this is too large for mapping purposes. Consider a $1°$ beam at a range of 50 km: the cross-range resolution is about 870 m. So our resolution cell at that range is very elongated: 1 m by 870 m. We have used pulse compression to reduce the range resolution to about 3 feet, but the cross-range resolution is still half a mile. To make good maps, we would like the resolution cells to be more or less square. Think of pixels in modern digital images; if they were very elongated then the pictures would look very strange and unnatural.

SAR enables us to reduce the cross-range resolution so that it is comparable to the range resolution. So, we can form radar maps with resolution cells—pixels—that are about 1 meter by 1 meter. Many remote sensors are coarser than this; for example, the satellite images of earth and other planets are obtained by SAR systems that are very distant from the surfaces that they are imaging, and the resolution cells are typically of the order of

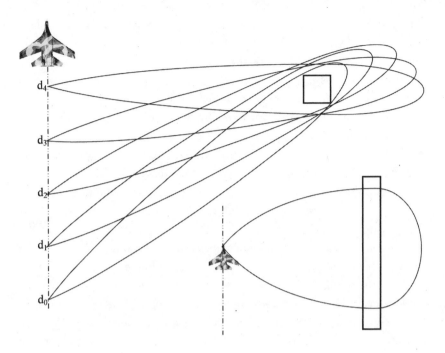

Figure 5.2 *Synthetic aperture radar. We want to image a scene (square box), so we point our radar beam at the scene as we fly by. In the simplest SAR version (below) we simply point our radar sideways and record the echoes for as long as the scene remains in the beam.*

10 meters by 10 meters. However, you see what I mean: SAR enables us to reduce cross-range resolution by a factor of 100 or 1,000, so that we can achieve square(ish) resolution cells suitable for forming maps. Now let us see how SAR works.

Synthetic Aperture

Consider figure 5.2. Here we have an airborne radar system that points its beam at a particular area on the ground, on one side of the aircraft. The beam direction is changed as the aircraft flies a distance of 1 km, say, so that the same patch of ground (the *scene*) is illuminated by the beam for the whole of that distance. Our radar transmitter sends out pulses at regular intervals t_0, t_1, t_2, . . . and receives the ground echoes. These echoes can be processed by a SAR to form a detailed map of the illuminated patch.

Look what is happening along the airplane flight path. The radar flies along a straight line and takes a snapshot of the scene every few centimeters. If the pulses are sent out at a repetition rate of f_r Hz, and the airplane speed is a constant $v = 200$ m s^{-1}, say, then the distance between snapshots is v/f_r—typically 20 cm. So the pulses are sent and received[3] when the plane is at location $d_{0,1,2\ldots}$ (fig. 5.2) where $d_{n+1} - d_n = v/f_r$. If the plane can keep the scene in the radar beam, while flying along a straight line for 1 km, then the radar processor will end up with lots of echoes. For any one pulse transmitted (say at t_{143}) there will be a bunch of echoes returned to the radar receiver an instant later, from different parts of the ground scene. Put another way—and this is the crux of SAR imaging—from any one part of the ground scene there will be echoes that return to the radar receiver at all points along the 1-km line. So the 1-km line can be thought of as a linear array of antennas, separated by v/f_r. To be sure, the echoes arrive at the "antenna elements" at different times than they would for a real linear array of this length but, if we can sort out the timing, then we have the potential for resolving the scene with the angular resolution of a 1-km antenna!

This is the essence of SAR. All the processing of SAR systems is connected with trying to sort out the timing; it is a significant problem for both hardware and software, but it can be done. (We do not want to end up with a Dali-like picture of the scene viewed from different perspectives.) Look at the benefits: a 1-km array will have an angular resolution of about $\delta\theta \approx \lambda/D$ (recall note T4), so for an X-band radar with antenna length $D = 1$ km this means $\delta\theta \approx 0.0018$ degrees, corresponding to a cross-range resolution of $R\delta\theta \approx 0.6$ m at 20 km range. In practice the achievable resolution would be a little broader than this theoretical limit, but a cross-range resolution of 1 meter could be achieved with this system.

Recall that the linear array achieves its resolution because of the way that the phases all add up; each element transmits a pulse (receives an echo) and the sum of these determines the overall transmitter (receiver) power in a particular direction. To achieve the narrow beamwidth of a linear array, that is to say high angular accuracy or resolution, the synthetic aperture of our SAR system must transmit and receive coherent pulses. The phase relationship between pulses is crucial to SAR image formation. Clearly the devil is in the details once more; if you want to lift the carpet to see what I have swept under it, please consult note T22. Here you will see some of the severe problems that must be solved be-

---

[3]There is a delay between pulse transmission and reception, of course, so the radar will have inched forward a little during this interval.

fore a focused image pops up on the SAR display. Timing details must be sorted. An image cannot be formed until the end of the 1-km flight, since all the data from all along the synthetic array have to be processed—so huge amounts of data have to be stored prior to processing. The airplane position and speed must be known *precisely*. For medium resolution SAR maps—say with 10-m pixels—the problems associated with focusing are less severe than for high-resolution maps. In the future, very high-resolution maps—10-cm pixels—will be attainable. Additional problems arise when trying to form SAR images at these resolutions. I do not mean only that the focusing has to be more accurate, but also that new sources of error arise. These sources are insignificant at lower resolution and so can be ignored, but they must be dealt with by additional processing when we attempt to construct very-high-resolution images.

Conventional wisdom tells us that the limiting resolution attainable by a remote sensor system is about the same as the wavelength. So, an optical camera will achieve higher resolution than a radar system because optical wavelengths are much smaller than microwave wavelengths. However, for SAR systems we may not be able to quite reach this one-wavelength limit. I believe that an X-band system, say with 3-cm wavelength, will not be able to achieve resolutions that are much better than about 10 cm, no matter how much processing we throw at the echo data. This is not because we lack the wits to understand what needs to be done, or because we lack the ever-increasing computing power required for higher resolution images, but rather because of one of the "new sources" of error that I mentioned above. Atmospheric turbulence, weather fronts, and other meteorological activity mean that the atmosphere through which our microwave pulses must travel is not uniform. Recall from chapter 4 that anomalous propagation results when waves travel through an inhomogeneous medium. Well, it turns out that we can calculate the likely effects of certain atmospheric inhomogeneities. For example, a weather front typically consists of a sharp transition in atmospheric density. At this front we therefore expect that microwaves will refract. Now suppose that a weather front crosses between our airborne SAR radar and the scene that it wants to image. The resultant distortions of the echoes result in image degradation. This degradation is small and insignificant for medium-resolution maps, but for very-high-resolution imaging it becomes a limiting factor. It cannot be compensated—in practice we will not know enough details about the weather front.[4]

Other sources of anomalous propagation will limit our ultimate SAR resolution capability before we reach the wavelength limit. Consider, for example, a city and, in particular, a factory with smokestacks. The atmosphere above the factory will be hotter and so less dense—our microwaves will bend when traveling through this zone. Other potential sharp-transition boundaries may contribute to anomalous propagation that bedevils very-high-resolution imaging, for example, at the edge of rain showers or clouds. These meteorological and other refractive effects will, in my opinion, ultimately limit our SAR resolution capabilities to about 3λ, rather than λ. We will achieve better than this on good days, but not consistently.

The SAR Menagerie

The SAR imaging parameters change as the geometry changes. So if, for example, our scene was 8 km away instead of 20 km, then the synthetic aperture needs to be "only" 400 m long instead of 1 km, to obtain the same 1-m cross-range resolution. If the scene is not roughly broadside-on to the platform, but off at an angle, then the parameters change again. A significant problem for SAR imaging is that it does not work in the forward direction. That is, straight ahead of (or straight behind) the airframe. Often, especially in a military context, this is the region of most interest—the pilot would like to know what he is heading into. Perhaps future generations will solve this problem or, more likely, they will learn from our masters how to do it. I say that because it appears bats have cracked this particular nut—it seems that they can form remote sensing images of objects directly in front of them. Not optical images, you understand, but acoustical images that permit them to see in the dark. But more of that later.

---

[4]In 2002 I presented a paper on this subject at the U.S. National Radar Conference, which was held that year in Long Beach, California. The conference was attended by very knowledgeable people from all over the country. (A number of us foreigners, who were deemed to have something interesting to say, also attended.) One group from a U.S. Air Force lab on the East Coast was interested in interferometric SAR (InSAR)—a very-high-resolution variant—and they asked me detailed questions about my paper. I was a little concerned, because my coauthor and I had assumed *very* sharp atmospheric transitions when performing our theoretical calculations, and I was unsure how realistic this assumption was. However, the InSAR folk told me that they had observed defocusing across several of their maps, which they had attributed to atmospheric effects, and which they could not eliminate. It seems that nature can throw up the kind of refractive effects that we were postulating, and that this does indeed seem to limit image quality.

The species of SAR imaging that I have discussed so far is called *Spotlight SAR*, for reasons apparent from figure 5.2. A simpler and more common species from the SAR menagerie is *Stripmap SAR*, and this is also illustrated in figure 5.2. Stripmap SAR produces images that are generally of lower resolution, but they require significantly less processing and are rather more robust—easier to focus. For Stripmap SAR the radar beam is broad and points in a fixed direction, straight out of the side of the airframe. The map is formed for a swath of ranges continuously. This is very useful for topographical mapping of unknown terrain—Venus, for example, or Antarctica. In a military or surveillance context, imaging radar will typically start off in Stripmap mode and then, if a particular section of the scene is interesting— a possible nuclear site in Iran, for example—the radar is switched to Spotlight mode for a better look.

Satellites are a natural platform for Stripmap SAR systems, because they confer several advantages. First, if they are in orbit over the earth, then the scene moves under them, they do not have to move over it. Second, the satellite motion is very smooth—there are no turbulence or cross-winds in space. This means that very long synthetic apertures can be processed, and consequently high-resolution images formed. Another advantage is that the satellite is looking down on the scene, rather than looking at it from a small grazing angle.[5] The major disadvantage of satellite SAR is the long range from platform to scene, so that even a very high angular resolution becomes a more modest cross-range resolution. Stripmap SAR can quickly image vast areas of territory. Consider the image of figure 5.3, obtained from the space shuttle, and you will get an idea of the capabilities of radar remote sensing.

I indicated earlier that we have to have synthetic apertures of different lengths if we are to image scenes at different ranges with the same cross-range resolution. This limits the range swath on which we can focus. DBS, for *Doppler beam sharpening*, is a SAR variant that does not have this problem. In DBS images the angular resolution is kept constant, and so the cross-range resolution is allowed to degrade (get bigger) as range in-

---

[5]A SAR platform at 10,000 feet altitude looking at a scene that is 30 miles (50 km) away will see it almost sideways. The microwave rays will slant at about 3° above horizontal. This grazing angle is typical for airborne SAR. Low grazing angle means that very hilly scenery will be largely in the radar shadow, where no image can be formed. It also means that the radar cross section of the scene is reduced, because the RCS is weaker at shallow angles than when viewed straight on. One advantage of modest shadowing is that it permits an estimate of scene feature height from the shadow length.

Figure 5.3 *A C/X-band synthetic aperture radar (SAR) image of San Francisco, taken from the space shuttle* Endeavour *in 1994. Note the Golden Gate Bridge* (left center) *at the opening of San Francisco Bay, and the San Andreas fault—the straight feature at lower left filled with linear reservoirs that appear dark. Note how the shadows help delineate topographical features such as hills. Courtesy NASA/JPL-Caltech.*

creases. In many applications this degradation is acceptable, and it permits us to image all ranges that are accessible to the SAR radar. The processing is also simpler. Not surprisingly, the resolutions that DBS images achieve are not as impressive as those obtained by the other SAR variants, at long range. There is a progression here: Spotlight/Stripmap/DBS corresponds to relatively high/medium/low resolution but also to small/medium/large maps.

A bizarre specimen in the SAR menagerie, one that is more distantly related to the rest—the duck-billed platypus of the SAR world—is *Inverse SAR* or ISAR.[6] Like the duck-billed platypus, ISAR is an opportunist and a specialist. ISAR is primarily a military radar mode and is used by the more sophisticated navies of the world to help classify enemy surface ships. Imagine that you (a Navy captain) have detected an enemy ship via conventional radar, and you want to find out what kind of ship it is. The sea is rough, the weather is foggy, with low clouds, so there is no point in sending out a helicopter for optical or IR imaging—they are blind in this weather. You don't want to get too close (to eyeball the ship) because he might be bigger than you are. So you send up a chopper with an ISAR radar mode to see if it can tell you something.

ISAR works—when it works—as a result of relative motion between the radar and the target, just as for SAR. Unlike SAR, however, the motion used by ISAR is that of the target, not of the radar platform. So, your chopper can hover at a safe distance, and point its radar antennas at the ship. If the ship is rolling on a rough sea, then that movement alone may be enough for the ISAR to work with. After ISAR has stared at the ship for a while, out pops a "movie" of the ship, on your radar screen.

Not a lifelike, crystal clear, and detailed color movie, but instead a grainy, fuzzy flip-chart type of movie. Worse, the movie does not fill in details of the ship's structure (portholes, colors, etc.); it just shows silhouettes. Nevertheless, such information is useful and can be sufficient for you to identify the type of enemy vessel you are dealing with. You can learn about the ship length, height, and outline; from the roll period (the length of time

---

[6]ISAR is the widely adopted (within the remote sensing field) acronym for inverse synthetic aperture radar, but there are others who vie for its use. You will find no further reference in these pages to the International Society for Antiviral Research, the International Society for Animal Rights, or to the International Symposium on Audit Research; nor will I have cause to mention again Intelligent Semi-Automated Reasoning, the Institute for the Study of Academic Racism, the Institut des Sciences Agronomique du Rwanda, the Institute for the Study of American Religion, or the German river Isar.

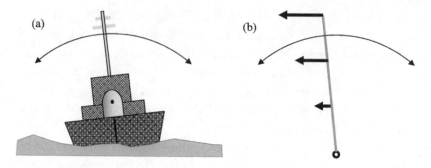

Figure 5.4 *Inverse synthetic aperture radar (ISAR). The rolling ship (a) has different speeds (b) depending on distance from the roll axis. So, echoes from different parts of the ship have different Doppler frequencies. This phenomenon can be exploited to form crude images of the ship.*

it takes to roll sideways and then right itself) you can say whether the ship is a destroyer or a cruiser, and so on. ISAR is opportunistic because it relies on the target ship motion, which is not under your control—if there is no rotating motion (roll is the most useful, since it is the biggest, but yaw and pitch may also be utilized by ISAR) then ISAR cannot form an image.

To see how ISAR works, consider figure 5.4. The radar echoes have been Doppler shifted because of the ship's rolling motion. For such motion, speed depends on height—the top of the mast is moving faster than the hull is. So, echo frequency depends on where the echo comes from.[7] If an echo comes from the top of the mast, then it will have a larger frequency than will an echo coming from part way up the mast. You can see how it might be possible to convert echo frequency back into height, and so plot an image of target height versus target range.[8] (High-range resolutions are obtained by the same methods as for conventional SAR imaging—for example, by chirp waveforms. ISAR provides the means for cross-range

---

[7]This fact—that the radar echo is shifted different amounts by a rotating object—has been exploited by some portable battlefield radars to identify helicopter targets. The helicopter rotors produce radar echoes that are characteristic of the chopper.

[8]Conventional SAR processing may also be explained in these terms—of target (scene) rotating relative to the radar platform—and this explanation would have emphasized the crucial role played by Doppler processing. However, I preferred the explanation in terms of synthetic beamwidth, because it may be a little more intuitively appealing. Those of you who plan on looking into remote sensing in more detail will come to realize that the two approaches are really two sides of the same coin.

resolution—here height.) In practice your radar will stare at the ship for about a quarter second before compiling an image. Then another quarter second will elapse before this image is updated. So you will see a sort of stuttering movie, showing an outline that seems to oscillate slowly back and forth, as the ship rolls. Destroyers roll once every 12 seconds or so, whereas carriers take 25 seconds. Destroyers roll up to 20° from equilibrium, in rough seas, whereas carriers roll less than 3° in the same seas. So, the movie helps you to classify the enemy ship.

You might imagine from my description of ISAR that it is a radar mode that is perhaps in its infancy, and that maybe one day the images will be much better, just as movie images have improved from grainy monochrome to clear color. You are probably right. ISAR (and SAR) have been improving ever since they first appeared, and there is no reason to think that this improvement will slow down any time soon.

Sidescan Sonar

Meanwhile, beneath the waves, *towfish* carry out the sonar equivalent of radar remote sensing. Towfish are linear arrays that are towed behind ships, at a controlled depth. The most sophisticated of these arrays form images via synthetic apertures—taking advantage of the array movement—but this is not necessary, nor is it so advantageous, for sonar imaging. First, the arrays usually move quite slowly, maybe only 3 m s^{-1}, not much more than walking pace, so it would take a long time to make a large synthetic aperture. Second, the real aperture beamwidths that can be achieved with high-frequency sonars are pretty impressive. Image formation from conventional signal processing (i.e., with real apertures) is much less complicated and so much less expensive. Third, the range of sonar remote sensors is much less than radar ranges, and so the cross-range resolution of sonar images is better. Let me illustrate sidescan sonar with some typical, representative numbers.

First, the name. *Sidescan* sonar towed arrays traditionally looked out sideways from the direction of motion, though nowadays the arrays can be steered in other directions. Very-high-resolution images are formed by transmitting very-high-frequency acoustic radiation—up to 1 MHz. For a towfish array that is 1 m long, this frequency provides an azimuth beamwidth of about a tenth of a degree. The elevation beamwidth is broad, maybe 40°, so that the array can illuminate a broad enough stretch of the ocean bed, as illustrated in figure 5.5. The main purpose of sonar remote sensing is to

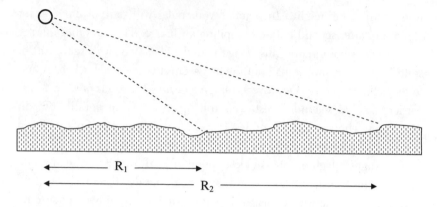

Figure 5.5 *A sidescan sonar towfish (circle) moves toward us out of the page. It ensonifies (beams acoustical waves in the direction of) the seabed between ranges R_1 and R_2. The azimuth beamwidth (in the plane of the paper) is paper thin: perhaps only 0.1°.*

map the seabed,[9] and we will assume that the geometry (fig. 5.5) is such that the sonar can receive echoes from the seafloor 100 m to 300 m away. The short ranges are dictated by the high frequency (short wavelength)— recall that acoustic radiation is attenuated by water more severely at high frequencies than at lower frequencies. At 100 m the cross-range resolution is 15 cm, or six inches. At 300 m it is three times as much. This is high enough resolution for most applications, so synthetic apertures are not normally required.

The towfish transmits one pulse, and waits until the echoes have been received from the maximum range of interest before transmitting the next pulse. As it is dragged along, the towfish acquires data that are processed to form an image of the seabed that is a continuous strip, usually of both sides (only one side is shown in fig. 5.5). Because towfish processing is simple, sidescan sonars are inexpensive relative to SAR systems, and because of this affordability sidescan sonars are widely available. You can readily Google any number of towfish manufacturers, who are only too

---

[9]This might simply be for topographical information, of interest to geologists and oceanographers. A common commercial reason for wanting an image of the seabed is to plan a route for an underwater cable or pipeline. Also, archeologists and treasure hunters use sidescan sonar to image the seabed, looking for sunken ships. Another commercial application, which I will say more about later, is for locating shoals of fish. One of the main military applications of high-resolution sonar imaging is for searching harbor entrances to locate enemy mines.

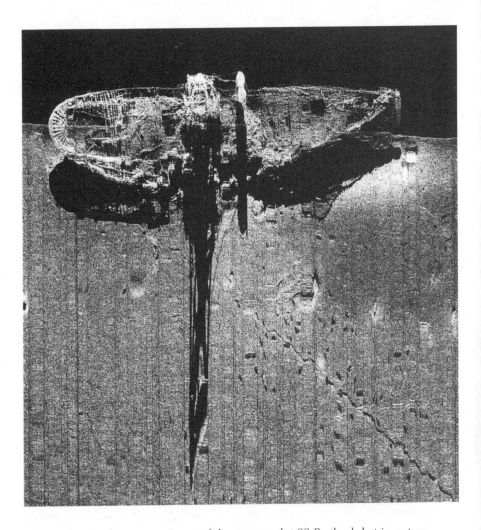

Figure 5.6 *A sidescan sonar image of the steam packet* SS Portland, *lost in a storm outside Boston in November 1898. Dark areas correspond to acoustical shadow—these enable the two-dimensional image to convey information about the third dimension. To the right are several large fish, which cast their own shadows. I am grateful to Bill Key of Klein Associates for permission to reproduce this image.*

pleased to show off their sidescan images. See, for example, the excellent sonar image reproduced in figure 5.6. This image looks like a photograph of a scene that is lit from the top, but it is the towfish transmitter "illuminating" the scene that gives rise to acoustical shadows—the plan-view image is the result of signal processing. The echo data are placed on a grid in their proper places (at the correct range and cross-range locations) and the echo power is color coded or, as here, converted to gray scale. No camera could see through muddy water to produce a photograph that is as clear as this acoustical image.

His Master's Voice: Enter MB888

Most bats speak through their mouths, as do most humans (politicians may be an exception . . .), but some bats speak through their noses. By "speak" I mean transmit echolocating signals, rather than communication signals.[10] Bats can form a directed beam of acoustical power (transmitter beam). The face of an echolocating bat has been adapted by nature for this purpose (fig. 5.7); for example, the peculiar leaf-shaped lobes on the noses of some bats are thought to help sound waves refract so as to concentrate the acoustical power in a chosen direction (with a beamwidth of roughly 60°). As with our sonar systems, these creatures listen out for the echo of their own voices, and interpret the results to provide themselves with information about their environment. In the words of one prominent bat researcher, Brock Fenton of the University of Western Ontario in Canada, bats use "the difference between what they say and what they hear" to collect information about their surroundings.

Bats are not the only creatures who utilize sound in this manner. Some birds (cave swiflets and oilbirds) and perhaps some pinnipeds (such as seals) transmit a series of clicks and interpret the results in a manner that constitutes crude echolocation. The best that these creatures can do is to learn about gross features of their environment that are not discernible optically, for example, the approximate distance to a cave wall. Odontocete cetaceans (toothed whales, including dolphins and porpoises) possess an advanced echolocation capability, which I will say more about later, but it is bats that have grabbed our imaginations as the icons of echolocation. This is due in part to the misconception that bats are blind—in fact, many species can see quite well—and so, the argument goes, the only remote sensing option that is available to bats is acoustical. This perception is reminiscent of the folklore about blind people's hearing being more acute than that of sighted peo-

---

[10]At this point I will note that bat *communication* sounds are quite distinct from their *echolocation* sounds. In general, the echolocation frequencies are higher and have a very different structure. Here we will ignore communication sounds, because what bats say to each other in the privacy of the night is none of our business. When they are hunting or navigating via echolocation, however, then they intend their sounds to be intercepted by others, and so we can listen in without pangs of guilt. If you would like to hear the output of bat detectors, they are available on several of the dedicated bat Web sites: for example, www.batcon.org/avscripts/script9.html.

Figure 5.7 *Pretty, he ain't, but the fleshy lobes around the nose of this furry flying signal processor make for an effective transmitter antenna. The ears are independently steerable receiver antennas. The binocular passive optical sensors are clearly an auxiliary feature in this MB800 model, a* Rhinolophus hildebrandti, *one of the horseshoe bats. This fellow is a constant frequency (CF) specialist. I thank Prof. Brock Fenton for supplying this photograph.*

ple.[11] The astonishing capabilities of some bats have impressed us humans ever since the 1930s when Donald Griffin (who coined the word *echolocation*) discovered that our distant mammalian relatives see in the dark by processing sounds.[12]

---

[11]According to Prof. Richard Dawkins of Oxford University, the uncanny ability of some blind people to detect an obstacle in their path is sometimes called "facial vision." What seems to be going on here is that blind people are unconsciously processing echoes (from the obstacle) of their own footsteps. If so, then some humans possess at least a crude echolocation capability.

[12]A large portion of the academic world was not very accepting of Griffin's work when it was first reported at a zoologists' conference in 1940. One indignant scientist grabbed Griffin's colleague Robert Galambos "by the shoulders and shook him while complaining that he could not possibly mean such an outrageous suggestion. Radar and sonar were still highly classified developments in military technology, and the notion that bats might do anything even remotely analogous to the latest triumphs of electronic engineering struck most people as not only implausible but emotionally repugnant." Recall that the Tizard technical mission from Britain to the United States took place in the same year.

Not all bats are equal. The 100 or so species of megachiropteran bats echolocate by producing a complex series of clicks, but their signals and the manner in which they process these signals is crude in comparison with the achievements of their diminutive cousins, the microchiropteran bats. There are 813 species of these little creatures, and all of them echolocate, making use of complex, structured tonal signals rather than simple clicks. They have developed many and varied echolocation adaptations, mostly in their (physically) tiny brains, vocal structures, and ears. The fact that all the microchiropteran bats have sophisticated echolocation faculties and aptitudes suggests that the crucial evolutionary adaptations arose in this group since it split off from the megachiropterans. Subsequent adaptive radiation has led to a flowering of many different echolocation strategies, which evolved independently in each species as these bats became specialized to different ecological niches and different prey—hence, the large number of microchiropteran species.

Tens of millions of years ago, all bats may have been diurnal fruit eaters. One theory posits that the evolution of raptors (birds of prey) pushed bats out of this niche and obliged them to become nocturnal. The bats developed two approaches to the problem of locating food in the dark. Megachiropteran bats specialized in night vision: detecting their food plants in low-light environments by sight. The microchiropterans switched to insect food. In particular, so this theory goes, they developed a taste for flying insects, and the means to acoustically detect and catch them. Whether or not this theory is correct, it is certainly true that microchiropteran bats had developed very sophisticated echolocation equipment as long ago as 50 million years.

The literature on the subject of bat remote sensing is huge and varied, encompassing acoustics, computing, military electronics, zoology, comparative physiology, neuroscience, and even philosophy. I do *not* propose to enter into any of these arenas in this book; bats are to be regarded here as clever airborne sonar systems covered in fur. I will peel back the fur, metaphorically speaking, to see what signal processing the bats do, and to make some guesses (some a racing certainty, others much more speculative) about how they do it. First, though, let us get some idea of the kind of strategies that bats adopt to find their way and locate their prey in the dark.

Microchiropteran Echolocation Strategies

To avoid wearing out the keys on my computer, I will henceforth ditch the label microchiropteran, and will simply refer to *bats*. Because I will now be restricting my attention entirely to the echolocation specialists, this economy should cause no confusion.

Figure 5.8 Eptesicus fuscus, *the big brown bat. This fellow transmits mostly descending-chirp FM pulses. I thank Prof. Brock Fenton for supplying this photograph.*

We can get a feel for the variation of bat echolocation strategies by considering the sounds that they produce. First, different species produce *very* different transmitter power outputs (in plain words, some are much louder than others). The variation between quiet whisperers (such as the northern long-eared bat and the vampire bat) and loud shouters (for example, the little brown bat and the big brown bat—fig. 5.8) is about 50 dB. The shouters really are loud but, fortunately for us and for most of the animal kingdom, the frequencies at which such bats shout is way above our hearing range. We would not get much sleep at night if shouting bats operated in the same acoustical frequency band as we do. Imagine clicks, chirps, and buzzes emanating from the night sky, each with the volume of a smoke alarm. Now multiply by a hundred (I'm guessing this is a reasonable ballpark figure for the number of bats within earshot) and you appreciate the potential problem.

We know a good deal about bat echolocation transmissions because of "bat detectors." These inexpensive devices are readily available, should you be interested in buying one. Bat detectors hear the bat sounds and then shift the frequencies down into our audible range, so that we can hear them. Batmen (as I shall dub the biologists and biophysicists of both sexes who research bat echolocation) have spent many years listening to bats forage in their natural habitats, and listening to bats that have been trained to participate in controlled laboratory experiments. From the large corpus of knowledge that they have accrued over the decades, and from our knowledge of signal processing, we can infer the target detection and imaging capabilities of bats. Later I will show you how batmen have "reverse engineered" the detected acoustical transmissions to gain an understanding of what signal processing the bats perform in their little (one-half gram—call it $1/60$ ounce) brains. For now, however, I will focus on what we have learned about bat-hunting strategies.

The shouters tend to operate in the open air, where there are few nearby reflectors (such as trees or the ground) that might contaminate the target echoes with significant clutter power. The targets of shouters are flying insects. Whisperers operate in more cluttered environments, such as woodlands or riverbanks. Here, loudness is a bad idea since the echo signals would be "clutter limited." (Doubling the transmitter output power would simply double the clutter power, and so the target signal-to-clutter ratio would not increase.) The transmitted waveforms are also different in different hunting scenarios. For airborne prey the bats usually emit a series of long, narrowband (i.e., narrow frequency range, known as constant-frequency, or *CF*) pulses or else a series of rapid, short clicks. For detecting insect prey that is hiding on a leaf within a tree, or sitting on the surface of a pond, more complex, wideband, frequency-modulated (*FM*) pulses are emitted.

Some bats (such as the horseshoe bats—fig. 5.7—and the Old World leaf-nosed bats) that specialize in flying insect prey can classify the target by its wing flutter. These bats emit the CF pulses. We will see how this classification works in the next section. Clearly the ability to classify targets confers a significant advantage, since the bat can reject those detected targets that are not edible, and concentrate its energies on worthwhile prey species from an early stage of the hunt. Other bats emit both CF and FM, depending on the stage of the pursuit. In general, there are three distinct phases to the echolocation attack sequence of a bat approaching airborne prey: detection, approach, and terminal. During the detection phase, many bats transmit long CF pulses at a low pulse repetition frequency. The approach phase pulses are typically shorter CF pulses transmitted at a higher rate. Terminal phase pulses are descending FM pulses of very short duration and high bandwidth, emitted at a high pulse repetition rate (the *feeding buzz*).

Bats can communicate over distances of fifty or a hundred meters, since communication signals are one way. They can use echolocation to navigate, detecting large topological features such as walls and riverbanks out to ranges of about 15 m. Bats probably remember such acoustic landmarks, and keep in their heads a map of the environment surrounding their roosts. At closer ranges, as we will see, their high-resolution echolocation capabilities enable them to avoid all environmental features in their path. The effective maximum echolocation range for hunting is shorter than for navigation, given the small insect echo cross section: only about 5 m. Once a target has been detected and classified as a potential meal, the bat locks on to the insect during the approach phase. When the distance between bat and insect (between interceptor and target) has been reduced to perhaps a meter, the attack sequence moves to the terminal phase. During this phase many bats form detailed images of the target (more on this later) and, if they decide to proceed to the kill, they intercept the insect. Airborne targets are trapped in a bat's cupped wings and then killed.

Bats often roost in large colonies, and they echolocate for navigation and hunting within earshot of hundreds of other bats doing the same thing. It is clear from observations that the bats do not get confused by the signals transmitted by others, even though these signals may be from conspecific bats with very similar calls. So, such bats must possess some sort of ATC (Air Traffic Control). I will discuss how humans achieve ATC in chapter 6; here we can only speculate about how bats achieve it. Apparently they can pick out the echoes that they transmitted from those that others transmitted—they recognize their own voices. Batmen have demonstrated this discrimination capability in controlled laboratory experiments, by playing loud artificial sounds, trying to jam the bat signal. However, a bat's ability to detect its own echo is very robust and it is not easily put off its stride. One trick that does fool a bat, however, is to play back a delayed recording of its own voice—it can be tricked into avoiding phantom obstacles that are not really present. Clearly, bats are picking their own voices out of a cacophony of others. Perhaps a bat colony allocates frequencies cooperatively between individuals, or perhaps individual bats adjust their calls until they find a form not currently used by others, so that they may recognize their own echoes. If the latter, then we may infer two features of bat signal processing. First, there is enough flexibility in the bat voice box and in the transmitted waveform to permit thousands of variations. Second, within the bat brain some very finely tuned filters exist that are capable of distinguishing between numerous echoes differing from one another in perhaps only a small detail.

In chapter 4 we learned something about the shenanigans that constitute EW. Most of these shenanigans are reflected eerily in the acoustical warfare between bats and their prey. Some insects can detect the calls emitted by bats that feed on them. For example, several moth species have evolved antennas that are the acoustical equivalent of radar warning receivers. Since they are listening to one-way transmitted signals rather than two-way echo signals, these moths can hear the bat echolocation calls at much longer ranges (40 m) than the bat can detect them. Upon hearing a predator signal, the moths can then adopt evasive tactics, such as seeking cover or moving away. The tiger moth, among others, can jam a bat's signal by transmitting noise at the right frequency whenever it detects a bat signal. In turn, bats have developed counter-countermeasures to get around these deceptive defenses.[13]

Bat Signal Processing

In this book we have explored several remote sensing features that were developed in the twentieth century to help us "see in the dark." Most of these developments had been anticipated (by some 50 million years) by bats: transmitter blanking, amplitude monopulse, sector scanning, integration, Doppler processing, correlation, frequency modulation, frequency agility, target classification, and image processing via ultra-high-resolution techniques. The last of these I will discuss separately, since it appears that bats are still way out in front of us; some of the more "conventional" standard signal processing techniques, as practiced by bats, are the subject of this section.

Transmitter Blanking

To avoid deafening itself, the echolocating bat must be able to shut off its ears while transmitting. It must then be able to switch them back on again very quickly to pick up the echoes from nearby objects. Special adaptations in their ears (well-developed muscles attached to the ear bones that transmit sounds) achieve transmitter blanking with superb timing—in some cases, the ears are shut off (muscles contract) and turned on (muscles relax) 50 times per second, synchronized perfectly with the transmitted pulses.

---

[13]Vision may perform a role here. A moth that puts all its defensive eggs in the "acoustical countermeasures" basket will not last long if the predator bat can *see* the moth. In human military terms, we refer to this combination of disparate remote sensors as *sensor fusion*; it is an increasingly important feature of modern military surveillance systems, as we will see.

Sector Scan and Amplitude Monopulse

Experiments show that many bats can achieve angular accuracy and resolution of about 1°. Yet their transmitter and receiver (nose or mouth and ears) beamwidths are much broader than this, about 60°. So, bats must employ signal processing to refine their estimates of azimuth angle and to separate targets that are close together in azimuth. Since bats have two ears, it seems sensible to suppose that they can take advantage of the horizontal separation between these receivers to employ an amplitude monopulse algorithm. The idea, you may recall from note T6, is to measure the ratio of echo power in each receiver and then, knowing the receiver beamshapes, to determine the angle from which the target echo arrived with greater accuracy. This method was applied in the CH radar. Computer simulations show that bats may very likely achieve the observed accuracy in azimuth estimation by this method.[14]

Improved elevation accuracy and azimuth resolution are more likely to be achieved by sector scanning. Horseshoe bats, for instance, have been observed to sweep their ears rapidly backward and forward while homing in on a target. Our experience of sector scanning in chapter 3 leads us to suspect that they are employing this technique. They are varying the beamshape across the target, just as required for sector scanning to work, and at a time when they require improved accuracy and resolution. Simulations show that the observed accuracy and resolution figures can be achieved in this way. Why else would horseshoe bats waggle their ears under such circumstances?

Doppler Processing

We know that bats employ some signal processing techniques, such as frequency agility, simply because these techniques can be observed directly. Other techniques, such as integration—to improve target detectability, you may recall—have to be inferred.[15] We have learned a considerable amount

---

[14]In my paper on bat signal processing (see the bibliography) I show that several common signal processing algorithms may explain how bats achieve the observed accuracies and resolutions. We cannot say for sure that they *do* use amplitude monopulse, but from the numbers that pop out we can infer that bats are more likely to adopt this method than others. In principle bats may also deduce the azimuth direction of a target echo by measuring the time delay difference, for example, or by measuring the difference in Doppler shift between the echoes in each ear. Simulations show that these alternatives are less robust and less accurate.

[15]From the range equation, and knowing the transmitter power of a bat and other range-equation parameters from measurements, we can calculate the detection range of a target. The observed detection ranges are greater, however, and so we infer that bats employ integration processing.

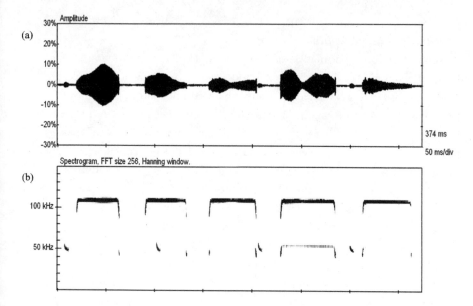

Figure 5.9 *Spectrogram of two bat echolocation signals, recorded together. The horizontal axis is time, covering an interval of 0.374 s. (a) The amplitude or power of the signals is represented by the width of the trace. (b) CF signal of a horseshoe bat (flat-topped pulses)—note the high duty cycle, that is, the large fraction of time spent transmitting. Also shown are the much shorter FM pulses (like inverted commas) that descend in frequency. I thank Prof. Brock Fenton for supplying this graph.*

---

about bat Doppler processing from both measurement and inference. Bat Doppler processing entails two significant strands, and these are associated with the two types of pulse: CF and FM. First, consider the Doppler processing of CF signals.

In figure 5.9 you can see a *spectrogram* of a typical transmitted CF pulse. Because of the long duration of CF pulses, and the unvarying frequency within each pulse, these waveforms have a very well defined frequency. So do the corresponding echo signals, though these echoes will in general be Doppler-shifted because of the relative movement of bat and target. CF bat hearing is particularly acute over a very narrow band of frequencies, an acoustical sweet spot, and there is a danger that the ever-changing movements of predator and prey might Doppler shift the echoes out of the sweet spot. But CF bats get around this potential problem: they adjust their transmitter frequency on the fly (literally) so that the echoes always hit the sweet spot, no matter what the bat–prey relative speed might be. Such dynamical adjustment is a difficult trick, but it pays great dividends, because the sweet

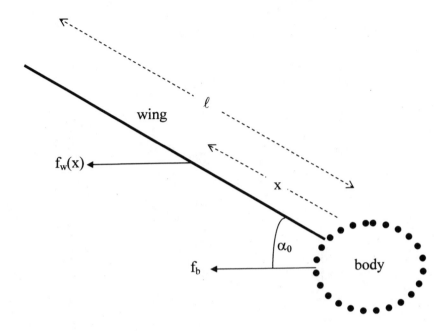

Figure 5.10 *Simple model of a moth, viewed from the front. Only one wing is shown. The bat echolocation pulse arrives from the left. The echo will depend in detail on moth parameters such as body and wing size, and wing beat amplitude (α_0) and frequency. The echo frequency $f_w(x)$ from the wing will depend on which part of the wing it comes from. The sum of echo contributions from all over the moth body determines the echo waveform of each pulse.*

spot is able to resolve very small frequency differences in the echo signal, and the bat uses this information to classify the target.

I can provide an appreciation of how this target classification process works with another computer simulation. Consider a very simple model of a flying insect target, say a moth, as sketched in figure 5.10. We can calculate the echo signal from such a target, assuming that the transmitted signal was CF. The detailed form of the echo signal will clearly depend on a number of target parameters: the cross section of moth wings and body, the wing length, the wing-flap amplitude, and the wing-flap frequency. In particular, the echo frequency will be spread because of the moth wing beats (recall the earlier examples of ISAR and of helicopter rotor echo characteristics). So, the echo pulses will have a wider bandwidth than the transmitted CF pulses. Doppler processing the sweet spot can tease out these frequency differences: the result looks like figure 5.11. I emphasize that my

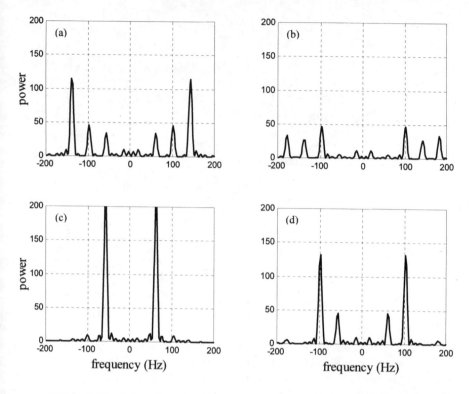

Figure 5.11 *Echo spectra for the moth model of figure 5.10, with different moth parameters. In all cases the moth wing-beat frequency is 20 Hz. The varied parameters are wing-beat amplitude (α_0) in degrees, wing length (ℓ) in centimeters, and the ratio (r) of wing to body echo cross section. (a) $(\alpha_0, \ell, r) = (30, 2, 4)$. (b) $(\alpha_0, \ell, r) = (30, 3, 3)$. (c) $(\alpha_0, \ell, r) = (20, 2, 4)$. (d) $(\alpha_0, \ell, r) = (20, 3, 3)$. The clear differences suggest that moths might be distinguished based on their different echo spectra.*

simulation is oversimplified: real moths are much more complicated targets than I assumed in figure 5.10, and the real echoes will look correspondingly more complicated than those of figure 5.11, but the model serves to show how CF bats can Doppler process echoes to classify targets. You can see that different target parameter values lead to very different echo spectra, and this spectral variation provides a basis for the bat to decide what the target is. A large moth will have a different spectrum from a small flying beetle.

The target model I used in figure 5.10 assumed that the moth was seen from the side, and that it did not change orientation, or any other parameters, during the echo process. Real moths will be seen at different angles, and these different aspects will produce different spectra. The aspect, and moth wing beat rate and other parameter values will change from pulse to pulse, as the moth changes direction or tries to elude the pursuing bat. These changes will all yield different spectra. Presumably, though, there is something in the quality of a given moth's echoes that transcend the differences, so that the bat can recognize the target as a moth, whatever it may be doing. In other words, the bat has in its brain a template of different target types, and can recognize from the template that a given echo falls within a certain category, just as we can recognize a face from different angles, or in different lights. There are significant implications here about the capabilities of bats' brains: the ability to *recognize* objects—indeed the ability to perceive objects as distinct entities and to classify two objects as being of the same or different type—goes a long way down the road of image perception. Such capabilities are usually associated with creatures that have much larger brains.

We reach a similar conclusion when considering the Doppler processing of FM signals. Bat FM signals are chirps, though not of the linear form that we discussed in chapter 3. Almost certainly, though, the chirp waveform has been adopted by bats for exactly the same reason as by humans: to improve range resolution. Recall from equation (2) that the resolution that can be achieved by a chirp waveform is $\delta R = c/2B$, where c is the speed of sound in air (for airborne sonar) and B is the chirp bandwidth. Without chirp processing, a bat's range resolution is determined by the pulse length. From figure 5.9 you see that the shortest of the FM pulses is at least 1 ms long and so the range resolution would be about 17 cm (say 7 inches) without chirp processing. With such processing, assuming a chirp bandwidth (from fig. 5.9) of 5–10 kHz we obtain a range resolution of 2–4 cm. This is enough to pick out one insect from a swarm. Some bats chirp with higher bandwidths and can attain higher range resolutions, down to about half a centimeter. Experiments with FM bats show that they can achieve resolutions that are much better than we would expect from the transmitted pulse lengths, and so we may safely conclude that FM bats use the chirp waveform to improve range resolution.

Bats that do emit chirp FM pulses have refined the technique in a very sophisticated way. You can see from figure 5.12 that the FM waveform is not *linear* FM, as we humans use in radar (chapter 3). It is *hyperbolic* FM,

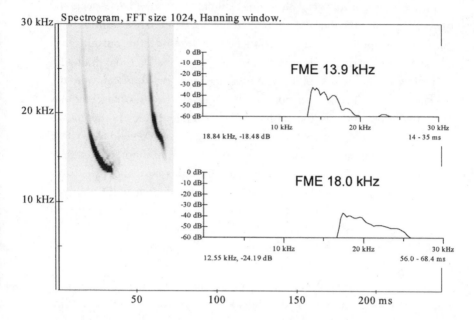

Figure 5.12   *Spectrogram of the pulse emitted by the FM bat* Tadarida teniotis. *Note the shape of the waveform* (left): *frequency does not fall with time in a linear manner, but the curve is instead a section of a hyperbola. This form fine-tunes the FM processing to extract optimum range resolution. I thank Prof. Brock Fenton for supplying this graph.*

and this is no accident; bats know their physics.[16] In note T15 we saw how the Doppler effect arose; if a single frequency f_0 is transmitted, then the echo frequency from a moving target is $f = f_0/\rho$, where $\rho = (c - v)/(c + v)$. (This is just another way of writing equation (3) of note T15.) Here v is the speed at which transmitter and target approach one another. What happens

---

[16]Not really, of course. All these sophisticated signal processing techniques evolved by natural selection of random mutations, so that the progress made by bats in evolving impressive echolocation capabilities arose empirically, over millions of years. This progression is *not* like the empirical progression of humans, who historically have developed technological techniques and designs (for example, in steel making or in bow and arrow design) that greatly exceeded their theoretical knowledge. Humans applied their intelligence to observe if a design change produced improvement, and quickly abandoned it if no improvement occurred. Bats evolved blindly and without a goal, directed only by the ruthless forces of natural selection that weeded out unsuccessful changes. Hyperbolic FM waveforms evolved simply because they are better than any other type in assisting bats' survival.

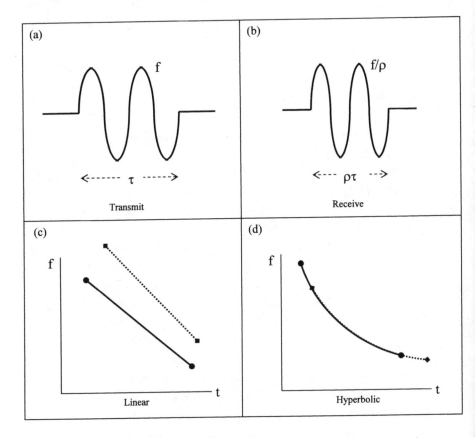

Figure 5.13 *Doppler squashing. A transmitted CF waveform (a) is compressed in duration (b), and so in frequency, by a factor ρ, where ρ is less than one. This is the Doppler effect. For a linear FM waveform (c) the slope of the line increases by a factor of $1/\rho^2$, so the transmitted and echo waveforms are different. The two waveforms do not correlate very well, and range resolution is degraded. For the hyperbolic FM waveform (d) the echo takes the same form, but is delayed in time. Correlation restored— range resolution optimum. For radar this does not matter since $\rho \approx 1$, but for sonar/echolocation it matters a lot.*

if the transmitted signal is a linear FM chirp, instead of a constant frequency? We saw in figure 3.10 that the chirp is delayed and the frequency is shifted. But this is only approximately true. In figure 5.13 you can see what really happens. For a linear FM transmitted waveform, the echo waveform is still linear FM, but the slope has changed due to the Doppler effect. We can ignore the difference in slope for radar, but not for sonar or echolocation, because the two slopes become more and more nearly parallel as c increases. The speed of light is much, much greater than the speed of sound in air or water, so for radar we can say that the echo waveform has the same slope—it is simply a shifted and delayed version of the transmitted waveform, as in figure 3.10. For echolocation, however, the echo waveform does not have the same slope, as you can see in figure 5.13. This fact spoils the signal processing of linear FM chirp signals. Correlation of the echo and transmitted signal no longer produce a sharp spike, which permits high-range resolution—the spike becomes smeared out, and range resolution is degraded. But bats have found a way around this problem.

Bats transmit the hyperbolic FM waveform. That is to say, the shape of the curve $f(t)$ is not a straight line, but is instead a hyperbola,[17] as shown in figure 5.13. Now the echo is truly just a shifted version of the transmitted waveform, without any change of form. So when the echo and transmitted pulses are correlated, they once again produce a sharp spike that permits the maximum range resolution ($\delta R = c/2B$). Computer simulation can also show that the accuracy, as well as the resolution, of range estimations is better for hyperbolic than for linear FM waveforms. Bats first thought of this ingenious solution to the shortcomings of linear FM; sonar engineers independently derived the same solution some 50 million years afterward.

Bat Image Formation: Hearing the Picture

The bandwidth of the moth spectra of figure 5.11 is a few hundred Hertz. The frequency resolution of the CF bats that process the moth echo signals is much less than this, and so some bats, at least, can form a reasonably detailed picture in the frequency domain. In fact, by looking at the moth with many different pulses, the bat can build up a picture of target movement. An idea of what this looks like for my simple moth model is shown in figure 5.14. To generate this figure I assumed that some of the moth parameters changed with time: the ratio of wing to body cross section, the speed of approach, and the wing beat rate. Real moth images of this type (power versus frequency versus time) will look *much* more complex, and will provide the bat with, potentially, a huge amount of information from which it can discern a lot about the target—in particular, whether it is tasty.

FM bats form spatial images, converting the wide-bandwidth echo pulses into very-high-range resolutions, thus building up a reasonable facsimile— a coarse-grained snapshot—of the target. Again, the bat can follow the ob-

---

[17]To be pedantic, it is a section of a hyperbola. If hyperbolas and you have never seen eye to eye, then don't worry about it; all you need to know about them is that they are just the right shape of curved line to solve the Doppler distortion problem.

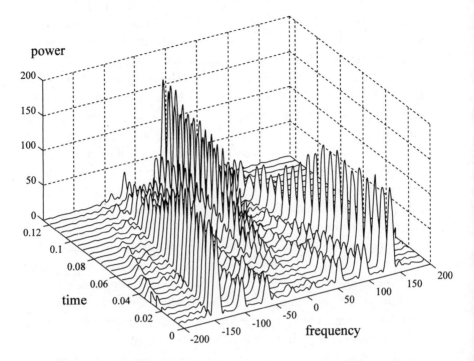

Figure 5.14 *Time evolution of the model moth spectrum (at time t = 0 the spectrum is that of fig. 5.11a). Here some of the moth parameters change linearly with time as the attack sequence develops.*

served features in time, pulse by pulse, forming a sort of moving sound picture of the target. It is tempting to say that bats are, during the terminal phase of their attack sequence, forming an image in their brains of the prey.

But maybe they do more than this. Or rather, they form better images than I have so far suggested. Experiments (perhaps controversial) within the past decade suggest that some bats can resolve objects that are separated by only 0.1–0.4 mm. This very-high-resolution capability provides a strong indication that these bats are indeed forming a detailed image of their immediate environment (what other purpose could such a capability serve?). There are two consequences of these new experimental results: one is philosophical and the other is technical. I will summarize only briefly the philosophical implications, since philosophy is not everyone's cup of tea, and will then quickly move on to the technical issues.

In 1974 the philosopher Thomas Nagel asked, in a well-known essay, "What is it like to be a bat?" He was referring to the ability of bats to interpret their environment in terms of sound pictures, a fact that was appreciated then and is more appreciated now. Philosophers are good at asking interesting questions but much less good at providing satisfactory answers. Nagel said that the images bats form will be unlike anything that we can imagine, because we are not bats. Thanks, Tom. Others suggest that the bats' images are akin to vision processing of optical signals. They form sound pictures that are different qualitatively from light pictures, but are processed as automatically and provide information (but different information) in as much detail. The images formed must incorporate the shapes of all the distinct objects perceived, so that they become segregated in perception, and these shapes must be updated with each pulse—a process that has been likened to computerized tomography. Not bad for a half-gram brain.

So much for philosophy. Now for the technical crunch: we don't know how bats can achieve such high resolutions. Batmen have measured the transmitted pulse lengths, and these are far too long to provide a 0.4-mm resolution. They have measured the bandwidths of FM chirps, and even the FM-ranging technique is ten times too coarse. If bats are indeed capable of these very high resolutions, then they must somehow combine the time-domain information with the frequency-domain information, and so they process a signal $S(t,f)$ rather than simply $S(t)$, as in conventional ranging, or $S(f)$, as in FM ranging. A crude indication may be envisaged via the undulating surface of figure 5.14. Perhaps a bat can "look" at a surface $S(t + \delta t, f)$ and see that it is different from the surface $S(t, f)$, for δt that is shorter than would be the case in either linear domain $S(t)$ or $S(f)$.[18] Who knows? Engineers would like to know, and the Office of Naval Research would like to know, it seems, since it funds much bat research.

Ultra-high-resolution capability has other ramifications:

- The number of resolution cells must be very large and so the processing load—the number of correlations and other calculations that must be performed to create an image of the scene—must be enormous.
- The manner in which bats transfer data between time and frequency domains cannot be the same as the way we do it (the *Fourier transform*), because our method only works if the spectrum

---

[18]Suppose the bat can just tell the difference between $S(t + \delta t_1, f_1)$ and $S(t, f)$, or between $S(t + \delta t_2)$ and $S(t)$. Then δt_1 is smaller than δt_2.

changes little in time. But bats see images with rapidly changing spectra.

- Bats do not form their images by anything like our SAR processing techniques; we know this because, you may recall, SAR does not work in the forward direction. Bats are imaging the insect prey right in front of them, so they have to be using an altogether different process.

Answers on a postcard please.

Cetacean Echolocation: Physics with a Porpoise

For this section I am going to use the phrase *whale biosonar* instead of the more precise but cumbersome *odontocete cetacean echolocation*. So here dolphins and porpoises are whales, for the sake of brevity. Also I have switched, for no very good reason, from ' "echolocation" to 'biosonar." These two words are synonyms, but here I prefer "biosonar" because it reminds us that whale acoustic remote sensing is akin to human sonar, since both operate underwater.

Arthur McBride was the first person to suggest, in 1947, that whales make use of biosonar. We now know that, compared with bats, whales are neophytes in this realm, since they have been at it for only 34 million years or so. In addition to biosonar, whales emit communication sounds, and some of this "singing" is quite beautiful. Please do not confuse whale biosonar with whale singing—the purposes and characteristics of these sounds are quite different. Much whale singing is infrasonic (the low frequencies travel greater distances, as we saw in chapter 4) whereas most whale biosonar is ultrasonic (the high frequencies lead to high-resolution imaging). There is great variation between species, but the general rule for whale biosonar is that the transmitted pulses are very loud and very short. The blue whale transmits biosonar pulses that are 80 dB louder than a Boeing 747. The bottlenose dolphin emits pulses that are only 50 *micro*seconds long. Such short pulses lead to a range resolution of a few centimeters, and this is plenty good enough for their purposes (orientation and fish classification). A typical burst repeats pulses every millisecond or so. The peak frequency of the pulses varies between species (between, say, 12 kHz and 200 kHz), and varies within individuals depending on circumstances. Because of the short pulse duration, there is only time for one or two wavelengths of the peak frequency within each pulse. This means that the pulses are rather broadband, and are modulated in amplitude more than in frequency, so in gen-

Figure 5.15 *There are few external signs that this dolphin can echolocate—streamlining takes precedence. The forehead bulge is the melon, which plays a role in transmitter beamforming. I am grateful to Dr. Simon Berrow of the Shannon Dolphin and Wildlife Foundation, Ireland, for permission to reproduce this photograph.*

eral we cannot classify them as either CF or FM. We will follow common practice and call all the whale biosonar pulses *clicks*.[19]

As with bats, whales have numerous biological adaptations to permit efficient acoustical remote sensing. They transmit biosonar pulses from either the nose or larynx, and these pulses are directed by either the melon or the skull (a consensus does not yet exist within the biosonar research community on this matter). The melon is a fat-filled cavity at the top of the head (see fig. 5.15). Other fat-filled cavities in the lower jaw receive the high-frequency biosonar sounds—the ears are sensitive only to lower-frequency

---

[19]Some dolphin whistles do constitute narrowband FM signals. The various biosonar noises emitted by whales have been called whistles, screams, grunts, trills, rasps, motor boats, and creaking doors. This diversity of descriptions emphasizes the sound variability. Individuals change the click characteristics depending on environment and purpose: increased bandwidth and power in regions with high ambient noise, reduced power in tanks and other environments with highly reflecting walls, reduced frequency and pulse repetition rate for detecting targets at long range. Low frequency and pulse repetition rate apply also when the whale is orientating itself, whereas higher values are used to classify targets.

sounds. It is possible that some dolphins can steer the biosonar beam without changing their spatial orientation. Also like bats, many whales have good (though monochromatic) vision.[20] Unlike bats, whales have large brains and are intelligent creatures. It is likely (though this is not yet universally agreed) that some dolphins are self-aware. We have seen that large brains are not necessary for high-throughput sophisticated remote sensing calculations, but they do help researchers estimate biosonar capabilities—because whales are easy to train.

Dolphin biosonar hearing is highly directional (high-gain receiver antennas, if you prefer): the receiver beamwidth is in the region 2°–4° for a narrowband sound source, and 0.7°–0.9° for a wideband source. When trying to classify a target, whales will change the frequency, bandwidth, and pulse repetition rate, rearrange the order of different pulses within a pulse train, and do everything they can to "illuminate" the target with different acoustical signals, probing for identifying features. Most research has concentrated on the biosonar capabilities of dolphins, because these creatures are easier to keep in captivity than their larger cousins, and because dolphins are friendly and cooperative. Here is a summary of what dolphin biosonar can do.

- Discriminate between two different fishes at a range of 15 meters.
- Discriminate between a piece of fish and an object of the same size and shape but made of different material.
- Distinguish one-centimeter-sized objects that are 15 meters away. This has been shown in tanks.
- Reliably classify hidden objects into one of two types. That is to say: object A or object B is hidden inside a water-filled box, which is then inserted into a dolphin tank. The dolphin senses the box, and can tell if it contains A or B.
- While blindfolded, identify individual people.
- Perform multi-echo processing, which suggests that they may have developed a synthetic aperture capability, and so can form high-resolution images, much as we do with sidescan sonars.

---

[20]Some specialists such as the river dolphins, who spend much of their time in murky waters, rely mostly on biosonar for their remote sensing. Most species, however, apply vision to supplement biosonar, or to replace biosonar in clear water. (It is thought that this is why dolphins get caught in fishing nets—they cannot see the nets, but could have detected them with biosonar, had it been switched on at the time.)

The available evidence suggests that dolphins perceive shapes directly, so that their biosonar images are functionally similar to visual images. It used to be considered that this was not so, but rather that dolphins used their intelligence to learn by association what an object might be, from its (abstract) echo signal. Now we know that they can perform all the usual remote sensing tasks (determine target range, direction, and speed) plus a few more (determine target size, shape, and internal structure). In fact, it seems that for dolphin biosonar the target internal structure is a most important classification aid. Perhaps a fish swim bladder resonates at some characteristic frequency, or the skeleton reflects acoustical radiation in a manner that differs between species.

Given these impressive remote sensing capabilities, you will not be surprised to learn that whale biosonar is of military interest. Coupled with their intelligence, dolphins are perceived by the U.S. Navy as potentially very effective minesweepers. The U.S. Navy marine mammal program (begun in 1960) is conducted by the Space and Naval Warfare Systems Center, San Diego.

BIBLIOGRAPHY

Several Web sites where SAR images can be viewed are available. See, for example, http://slisweb.lis.wisc.edu/~heather/653/sar.html.

A technical review of SAR processing, a very detailed pedagogical survey for the dedicated student, can be found in the classic paper *Developments in Radar Imaging*, by D. A. Ausherman, A. Kozma, J. L. Walker, H. M. Jones, and E. C. Poggio, *IEEE Transactions on Aerospace and Electronic Systems* **AES-20** (1984) pp. 363–400.

I can find no instructive account of ISAR that is intended for nonspecialists. During my time at a multinational aerospace corporation I wrote a terse in-house report entitled *Idiots Guide to ISAR Design* that was intended for managers, but this was very dry. The problem in recommending specialist literature is the opposite: there are many technical expositions to choose from. An excellent introduction to this much-studied topic is the venerable paper by J. L. Walker "Range-Doppler Imaging of Rotating Objects." *IEEE Transactions on Aerospace and Electronic Systems* **AES-16** (1980) pp. 23–52.

Bat echolocation literature abounds, at all levels of presentation. Two accessible and informative articles are those of M. B. Fenton: "Seeing in the Dark." *Bats* **9** (1991) pp. 9–13 and "Eavesdropping on the Echolocation and Social Calls of Bats." *Mammal Review* **33** (2003) pp. 193–204. The second article is available online, with many cross-references. For a physicist's account, see my review paper "The Physics of Bat Echolocation: Signal processing Techniques." *American Journal of Physics* **72** (2004) pp. 1465–1477. A very readable introduction to bat echolocation can be found in chapter 2 of R. Dawkins, *The Blind Watchmaker*. Penguin, London, 1988.

Several very useful biosonar Web sites containing educational information, including marine mammal sounds, are available. You can listen to various cetacean sounds at www.cetaceanresearch.com. The U.S. Navy marine mammal program is at www.spawar.navy.mil/sandiego/technology/mammals/.

6 ....  Special Applications & Advanced Techniques

We have explored most of the extensive territory that constitutes remote sensing. We have trekked through a morass of technical notes to reach the lofty peaks of signal correlation, Doppler processing, and synthetic apertures. I hope that you have gained an appreciation of the lay of the land. If you wish to explore in more detail, I have referred to other guides that will take you into whatever particular valley or hilltop appeals to you. A few off-shore islands remain that I will expand upon in the first part of this chapter, because they represent important applications of our subject, hitherto unexplored, that everyone who claims an interest in remote sensing should know something about. These special applications have evolved into unique and sometimes strange forms that reflect their unique realms. Once I have delineated weather radar, air-traffic control (ATC), fish-finding sonar, and echocardiograms, I will devote the second part of this chapter to an overview of some advanced techniques that, generally speaking, are still on the drawing board or at the experimental stage. When fully developed these techniques promise to continue the improvement of our remote sensing capabilities.

Weather Radar: Looking for Clutter

Every day on the TV weather forecasts we see large maps of the earth's surface, superimposed with atmospheric pressure contours, cloud cover, and meteorologists' graphics. These maps are constructed from a wide variety of sources (data from weather balloons, ground stations, satellite images, and radar) compiled during the preceding hours, and present weather forecasts that have been calculated by large number-crunching computers. The physics of weather formation and development is fairly well understood, enabling meteorologists to extrapolate the current measured state of the world's weather for a day or two into the future. This is what you see on the TV forecasts.

My local TV weather channel predicts sunshine, rainfall rate, and temperatures up to two weeks ahead, but this is a joke; meteorology is incapable of accurate predictions more than a few days into the future because of the well-known "butterfly effect." Atmospheric physics is governed by equations with solutions that are *chaotic*. This technical term means that, starting with two very slightly different sets of meteorological conditions at, say, midday Tuesday, the equations predict somewhat different weather for midday Wednesday, quite different weather for midday Thursday, very different weather for Friday, and so on. In other words, chaotic systems are characterized by equation solutions that critically depend on the initial conditions. The problem is that we do not know the "initial" conditions of the weather (say at midday Tuesday) since we cannot measure them all. Think of what such measurement would entail: we would need to know the temperature and pressure everywhere in the atmosphere all over the earth at that time, as well as humidity, wind speed, cloud formation, etc. We would also need to know the surface and subsurface conditions of all the world's oceans, because the interaction of atmosphere with the oceans influences weather significantly. Obviously, our data gathering of meteorological information is incomplete and imperfectly measured, so our knowledge of the state of the weather at midday Tuesday is only approximate. But the chaotic equations of meteorology require *exact* knowledge to predict reliably into the distant future. Any errors—say by not including the effects of a butterfly flapping its wings somewhere in Asia—and our predictions for weather in Baltimore a week from Friday will be wrong.

Good weather forecasting requires good weather measurement. The better we can estimate the current conditions over the earth, the further into the future our weather predictions will be (approximately) valid. This is where weather radar comes into its own: radars provide a great deal of information about the meteorological state of the atmosphere, in a very short time. Weather radars transmit microwave radiation of wavelengths between about 1 and 10 centimeters. The short wavelengths are better at picking out

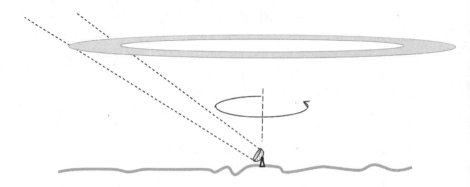

Figure 6.1 *To produce a plan position indicator (PPI) image, this weather radar scans at constant elevation angle. Strong reflectors at a certain altitude show up as a bright band. The reflector altitude is easily calculated from the band radius.*

small cloud droplets or drizzle, they provide better angular resolution, and, because short-wavelength radars are generally smaller, they are much less expensive. On the other hand, longer microwave wavelengths penetrate better through the atmosphere—they suffer much less attenuation and so operate at much greater range. Doppler radars detect objects in the atmosphere such as raindrops (or swarms of bugs) that are carried along by the wind, and so tell us about wind speed. Some atmospheric conditions such as turbulence result in significant changes in atmospheric density over short distances, comparable to the radar wavelengths; when this occurs, diffraction effects (see note T21) lead to radar pulses bouncing directly off the air, sending echoes back into the radar receiver.

Doppler weather radar displays tell meteorologists about the location of precipitation (map reference), altitude, type (rain, snow, or hail), precipitation rate, wind velocity profile (variation with altitude and with location), and cloud height and coverage. From this information the meteorologist can observe directly, or can infer, the location and movement of weather fronts and cells. To provide you with a simple example of how weather radar yields such information, consider figure 6.1. Here we have a narrow-beamwidth meteorological radar with an antenna that rotates in azimuth while maintaining a fixed elevation angle. Once it has made a full 360° rotation it will then change elevation to cover a different part of the sky for the next rotation. In this way a large volume of atmosphere is covered. Any precipitation within the beam sends (Doppler-shifted) echoes back to the radar receiver, so that the radial velocity of the precipitation can be plotted on a plan position indicator (PPI) map, producing a *velocity field* image. These images consist of arrows plotted on a plan-view map, with arrow direction indicating wind direction and arrow length indicating wind speed, or else speed and direction may be color-coded, so that the map appears as a color pattern. Whichever method of display is favored, the meteorologist can tell from the pattern how wind speed changes with altitude and with location.

The weather radar can also produce *reflectivity field* images, which plot the echo power as a function of location — just like the clutter maps or side-scan sonar images we saw earlier, only now the map is of the atmosphere. From differences in reflectivity (echo power) the meteorologist can infer rainfall rate or precipitation type. He may also be able to infer the altitude at which the atmospheric temperature reaches the melting point of ice; this is shown in figure 6.1. Water usually precipitates out of the air at high altitudes where the temperature is below freezing, so that the precipitation is as snow or hail. Atmospheric temperature is higher nearer the earth's surface, and so our snowflake heats up as it falls. At the altitude where water melts, the snowflake develops a skin of water, and water is much more reflective (+9.5 dB) than snow. So, at this altitude the precipitation produces stronger radar echoes. As the wet snowflake falls further it melts, forming an aerodynamic rain droplet that falls faster. Thus, at these lower altitudes the density of precipitation falls as it changes into rain. These changes cause a reduction in radar echo. The total effect is thus of enhanced radar echo at a narrow band of altitudes where the snowflake is melting. For our scanning weather radar, this produces a donut-shaped ring of strong reflections at the melting altitude. This is known as a *bright band*, and is sketched in figure 6.1.

So, you can see how meteorologists use a mixture of knowledge about atmospheric physics and radar systems to learn about the atmosphere by radar remote sensing. Of course, there is a lot more to radar meteorology than I have sketched here, but I hope you have gained a feeling for the role of radar in this field. Radar provides information about a large volume of the atmosphere, and from the patterns produced on PPI displays, meteorologists have learned to infer different wind profiles, and so on. Other antenna scanning modes (for example, "nodding" in elevation while maintaining constant azimuth angle) provide further information about sectors of interest. Other meteorological phenomena, such as tornadoes, have their own characteristic shapes on PPI displays, and so forth. In the future, higher-resolution radars will help to provide more detailed and accurate estimates of meteorological conditions at a particular instant, and from this better-known initial state the number crunchers will produce better forecasts, which will reach further into the future.

ATC Radar: Targets That Want to Be Seen

Earlier I promised that I would provide an explanation of Air Traffic Control radar operation. Like most remote sensing applications, it began in World War II. Britain was the center of a lot of air activity at this time, with many hundreds of U.K., U.S., and German planes buzzing around the island. The RAF introduced a remote sensor system that satisfied the fundamental military need to differentiate between friendly and enemy planes: Identify Friend or Foe. A friendly plane would intercept radar pulses—much as a radar warning receiver does—and respond with coded pulses. IFF systems are still in use today, only they are much more sophisticated because the airspaces are much more crowded. IFF was adapted to the nascent civilian airline-traffic control systems shortly after the war. The one-way nature of IFF beacons means, of course, that relatively inexpensive, low-power units could transmit a long distance.

In fact, ATC began shortly before the war. The first ATC tower was built at Newark International Airport in 1935. By using maps and radio communications with the approaching planes, the tower could plot the position of nearby airliners. A decade later LaGuardia Airport added radars to track planes—the first civilian application of radar—and thereby increased the landing rate from five to fifteen aircraft per hour.

Since the 1950s, ATC has grown exponentially, in delayed response to the exponential growth in civilian air traffic. Sadly, the spur for each stage of growth has often been an air crash, or a fatal collision between two airborne planes. The Federal Aviation Agency (later the Federal Aviation Administration) was born out of such an accident, and was tasked with implementing a nationwide ATC system that would maintain a safe separation between all airborne commercial aircraft in U.S. airspace at all stages of each flight. Over the past half-century the FAA and equivalent overseas administrations have struggled mightily to achieve air safety worldwide. You can judge their success from the statistics graphed in figure 6.2. There is no clear indication that, worldwide, the number of fatalities per year is falling. However, the density of air traffic has greatly expanded and, as you can see, the number of accidents per million

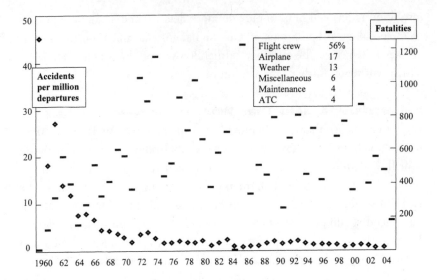

Figure 6.2 *Commercial airline accidents (diamonds) and fatalities (bars) per year.* (Inset) *"Hull loss" accidents broken down by cause. Data from* Statistical Summary of Commercial Jet Airplane Accidents. *Boeing Commercial Airplanes, May 2005.*

aircraft departures[1] has fallen markedly. Also shown is the cause of air accidents, for flights where this is known. Note that ATC control errors account for only a very small portion of accidents.

Not shown in figure 6.2 is the variation in accidents throughout the world. North America accounts for no less than 72 percent of flights, but only for 18 percent of air accidents, a ratio of 0.25. The equivalent ratio for Europe is 1.5, for Asia and Australasia it is 3.1, for Africa and the Middle East it is 3.4, and for South America is 9.0. Another interesting breakdown of the statistics is by flight stage. Most accidents occur during the airplane final approach and landing. The next most accident-prone stage is takeoff. The deadliest stage, however, is during the relatively brief climb, which accounts for 25 percent of fatalities. It is not difficult to understand these statistics. Accidents are common at takeoff and landing due to stresses placed on the airframes and to air traffic density. Because of nearby emergency services,

---

[1]The number of plane departures is not quite the same as the number of flights, since a few of the planes do not arrive, due to accidents. As you might imagine, the statistical analysis of civilian airliner accidents is an industry in itself and has produced some very detailed and interesting results. See the Boeing reference cited in the caption to figure 6.2 for further analysis.

however, these accidents result in fewer fatalities than occur during the climb stage. This stage places strain on an airframe, when it is loaded with fuel, and perhaps while it is over the sea or in open country distant from prompt assistance.

During the 1950s, ATC centered on rules and manual procedures. Passenger planes approaching an airport runway followed predetermined flight corridors and were booked in to land at specific times in a specific order. Radar provided information to the airport control tower about plane direction and range. The story since then has been one of struggle against ever-increasing traffic density. Following a collision between two passenger planes over New York in 1960, the ATCRBS secondary surveillance system was created. The Air Traffic Control Radar Beacon System utilized primary radars for locating and tracking aircraft over U.S. airspace, and secondary radars for aircraft near airports. These provided information about aircraft altitude as well as direction and range. Transponders on-board planes transmitted details about aircraft type and flight plan. This system plugged a gap, but by the late 1960s it was clearly overloaded. One problem was that the airport radars had rotating antennas—necessary to cover the whole area and to provide frequent updates of each plane's location. But each update caused each transponder to generate a new signal to the airport control tower, producing an overload due to unnecessary multiple responses. This would not be a problem today, but the computer facilities in the 1960s were too primitive to identify and eliminate these spurious responses.

The Traffic Collision Avoidance System slowly replaced ATCRBS and operates today. TCAS provides more accurate information about the location of each aircraft (multiple receivers at airports triangulate aircraft location through precise timing of aircraft signals) and is much more integrated with other ATC systems around the world. Aircraft can now approach, land on, and exit a runway within one to four minutes, enabling a throughput—allowing for departures—of 30 planes per runway per hour at a busy airport. The "holding patterns" so common in the 1960s, whereby planes approaching a congested airport were obliged to stack up in layers overhead until a landing slot became available, are now uncommon. It is safer to delay a plane's takeoff, or have it slow down en route, so that it arrives when a slot is free. ATC has become very specialized. There are ground controllers, local (tower) controllers, departure controllers, center controllers, and radar controllers, each allocated a specific task dealing with aircraft at a specific location or stage of flight. The world is divided into many ATC zones, with handover from one region of ATC control to another as an aircraft crosses boundaries. *Terminal control* guides aircraft that are within a 50-mile range and 10,000 feet in altitude of an airport; *en route control* guides aircraft between terminals. The sum of terminal and en route air spaces over the earth constitutes *controlled* air space; the rest is *uncontrolled*. The United States is divided into 21 ATC centers, each of which is subdivided into sectors. Each sector is the responsibility of a controller. Icons on large display screens provide information about aircraft flight plan, speed, and location. Software anticipates potential collisions and recommends collision avoidance maneuvers based on the known aircraft location, heading, speed, and climb rate. Integration of data from primary and secondary radars and transponders provides the ground controllers with "situational awareness" and provides the pilots with what they need to know about nearby aircraft.

The extensive ATC network is very effective, but is under strain. Traditional sources of problems are ground clutter (strong clutter reflections from structures near the primary radar sites, which sneak into the radar receivers through the antenna sidelobes), weather, and traffic density, of course. The first source is pretty much under control. Weather causes headaches, for example, by closing airports or requiring airplanes to maneuver around thunderstorms, so disrupting carefully prepared approach routes. The integration of worldwide ATC helps avoid congestion and other weather disruption through increased forewarning.

Future ATC may not be ground based at all. Airborne planes will message each other with all the information they need; precise locations may be provided by GPS. This fundamental change in our approach to ATC means that potentially there will be no uncontrolled air space, because all aircraft will carry around their own ATC.

Fishy Sonar

At a sonar conference in the mid-1980s I met a Frenchman who was researching fish-finding sonars. These devices are economically important and widely used, though it is unusual to find someone who works in this field at a military sonar conference. He must have felt like, well, a fish out of water since most of the rest of us regarded his targets as clutter (recall the copulating shrimp and farting herrings). We concerned ourselves with sonobuoy deployment, and he with fishes' swim bladders. We were looking for Soviet subs and he was looking for herring.

All commercial fish-finding sonars are active; that is, they transmit pulses and listen for the echoes. The sonar antennas may take the form of large towfish linear arrays with a range of several kilometers, or they may be simply handheld devices that are placed underwater to search out a single fish a few meters away. Most fish-detecting sonars are intermediate in scale and complexity, and lie between these two extremes in expense. They are often housed within a boat, say a commercial trawler, with through-hull transducers (acoustical transmitters). Whatever form the sonars take, the processed echoes are displayed on a screen. All this sounds like conventional sonars, which after all is why the Frenchman attended our conference. However, there are significant differences in the postdetection processing, at least for the more sophisticated devices.

Simple commercial fish-finding sonar can detect schools of fish, and the shape of these schools. The schools move against a fixed background, and so can be identified as fish. The sonar will be able to provide information about the direction, range, and depth of the school, and provide some indication as to its size. More sophisticated devices transmit pulses (at a low pulse repetition rate, to avoid range ambiguity—explained in note T23) with a linear FM waveform. This waveform is utilized not to improve range resolution, but to distinguish between different species of fish. Experiments show that the structure of the echo waveform varies with target species in a manner that is consistent and reproducible; in other words the echo shape depends upon the fish species. Cod, pollack, mullet, and sea bass, for instance, all have different swim-bladder shape, size, and orientation. It is the air-filled swim bladders that cause strong echoes, in particular, at certain acoustical wavelengths, because they resonate like organ pipes. The acoustical pulses likely bounce around within the fish (between swim bladders and skeletons, for example) before echoing back to the sonar receiver. These multipath effects will account for the complex species-dependent structure of the echo waveforms.

Most commercial fishermen rely on fish-finding sonar to locate their prey. (The sonar can double up as a depth sounder.) Their sonars have a range of a few kilometers, can identify fish species, and estimate school size, location, and heading. The sonars can readily identify marine mammals by their characteristic echoes (I guess lungs look different from swim

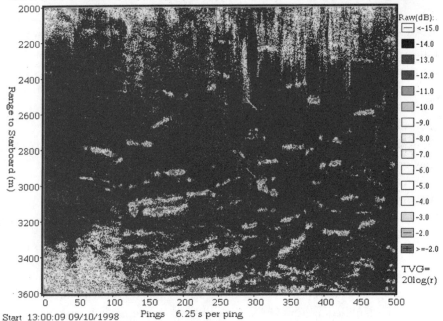

Figure 6.3   *An experimental towfish sonar display. The vertical axis shows range from the towfish, which is being towed from left to right over a distance of 6 km. Bright blobs represent schools of fish, probably herring. IRFS, Intermediate Range Fisheries Sonar; IOS, Institute for Ocean Sciences; TVG, Time-Varying Gain. I am grateful to Svein Vagle of the Institute for Ocean Sciences, in British Columbia, Canada, for permission to reproduce this image.*

bladders), and can mitigate the problem of ship noise—a significant source of interference—by signal processing. Figure 6.3 shows a fish-finding sonar display.

Fish-finding sonars have been put to use in trying to locate the Loch Ness monster. Nessie still eludes us, however. I suppose it would be churlish of me, and disloyal to the Scottish Tourist Board, to suggest that Nessie's shyness is due more to her mythological nature than to the inadequacies of fish-finding sonar.

Echocardiology

I have a confession to make. I have revealed my heart to a lady who is not my wife. In fact, to two ladies, one a Chinese physician and the other her Antipodean student. However, fret not for my moral lapses; these ladies were seeing real-time sound pictures of my beating heart, on a cathode-ray tube (CRT) monitor, seeking to understand the source of a cardiac arrhythmia. Medical imaging is a large part of remote sensing nowadays. The regions being imaged are far from remote, but they are inaccessible; the alternatives to imaging are invasive exploratory procedures that are more costly, more risky, and less informative. Of the many types of medical imaging in use today (computed tomography [CT] scans, X-rays, and so on), I choose to outline medical ultrasound imaging—in particular, echocardiograms—because they are closest in spirit to the traditional active sensors that we have examined in the rest of this book.

The ultrasound transducers operate at very high frequencies—between one and twenty MHz—and, by processing the echo signals in the manner described in earlier chapters, we can obtain range resolutions of a few wavelengths. The speed of sound inside the body is somewhat variable, depending on tissue density, but it is typically about 1,540 m s^{-1}. (This is only slightly greater than the speed of sound in water.) Thus, an echocardiogram transducer frequency of 3 MHz can produce images with $\delta R < 1$ mm. The angular resolution is not particularly impressive, perhaps $10°$, but this does not matter much since the range is so short—only a few centimeters (so echocardiology has the shortest range of the remote sensing applications described in this book). There are two types of echocardiogram: the regular *transthoracic* and the occasional *transesophageal*. For my transthoracic echocardiogram, I had some Jell-O-like goo smeared over the left side of my chest to form good acoustical contact with the transducer, which is a small handheld device placed carefully by the technician (a *sonographer*). For transesophageal echocardiograms the transducer is placed down the throat, and so ensonifies the heart from behind. There is less obstruction in this case (chest wall, and so on, between transducer and heart), and transesophageal echocardiograms are required for obese people or those suffering from emphysema.

The images formed by echocardiograms are similar wherever the transducer is placed. One type of image presents on the display a two-dimensional slice through the heart, which shows in real time the size, thickness, and movement of the chambers, valves, and blood vessels in a beating heart. Another type of image (or one superimposed on the first type) displays Doppler information—color-coded displays of blood velocity at different places in the heart. (See figure 6.4 for two examples of echocardiograms.) An audio signal is also played—a sloshing sound—that conveys information to the technician if not to the patient.

Used in conjunction with an electrocardiogram, which records the timing of electrical cardiac events (the pumping cycle), echocardiograms convey a great deal of information. From the observed size of heart chambers, cardiologists can learn about diseases of the heart muscles and high blood pressure. They can observe the pumping function and estimate the heart's *ejection fraction*. EF is the fraction of blood that is pumped through the left ventricle of the heart during one beat. It is normally in the range 60 to 75 percent for men (a little lower for women), but if a heart is not firing on all cylinders then the EF falls. From EF a cardiologist can infer cardiomyopathy or aneurysm. Echocardiograms also reveal how well the heart valves function; the structure, thickness, and movement can signify tissue that has become scarred through infection or has become calcified or torn. Doppler processing can reveal abnormal backleakage through the valves. Any abnormal structures present (blood clots or tumors) are visible on an echocardiogram.

Echocardiograms are noninvasive and safe—side effects are rare—and they image muscle and other soft tissue very well. The sonographer sees the heart as it usually functions, and so the act of observing does not significantly influence the observation. Thus echocardiography provides an unbiased view of heart function. Another advantage of echocardiogram images is that they are relatively inexpensive. On the negative side, ultrasound does not penetrate very deeply. It performs poorly when air (in the lungs, or between transducer and skin) gets in the way. Bones and other tissue of different density cause multiple reflections and interference. Because of these sources of variability, sonographers need to be highly trained. Even so, the interpretation of echocardiogram images depends to some extent on the person who conducts the test. My sonographer moved the transducer and me every which way to find a good position for imaging.

Echocardiograms were first invented in Sweden, and independently in Scotland, in the 1950s. During the 1960s they spread as diagnostic tools, and have come a long way since then. Increasing computing power permitted increasingly sophisticated signal processing algorithms to be applied to raw echo data, and we can look forward to this improvement continuing in future decades.

Advanced Techniques: Bistatic Radar
The word *bistatic* is formed from roots meaning "two" and "stationary," which is unfortunate since bistatic radar can refer to "many" and "moving." The

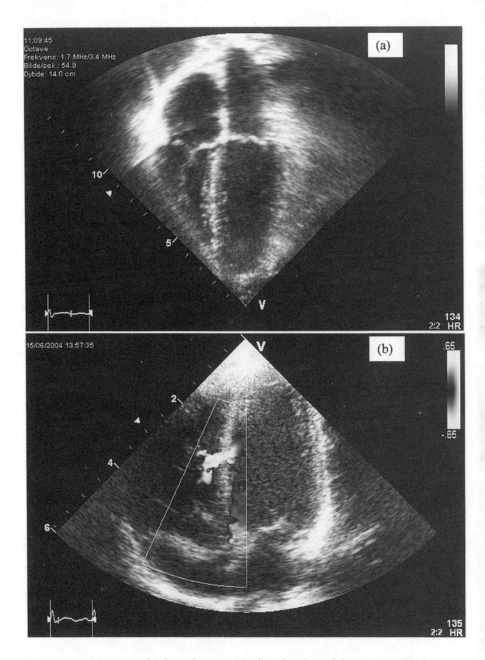

Figure 6.4 *(a) A normal echocardiogram. The four chambers of the heart are clearly visible. (b) An abnormal echocardiogram. The image shows "a mid-muscular ventricular septal defect, typical position and size found in a new-born child." The trace at bottom left shows the cardiac cycle, with a line indicating the time within the cycle. Images provided by Norwegian sonographer Kjetil Lenes.*

idea behind bistatic systems is that the radar transmitter and receiver(s) are not colocated; they may be separated by distances that are comparable to the target range. There are significant advantages to this arrangement that I will describe—and then give you some idea about the problems entailed in realizing this spatially distributed form of radar operation. Here I will follow standard practice and refer in general to all species of radar with separate transmitters/receivers as *bistatic*, and will use the same term more specifically and more correctly for two stationary separated antennas (one transmitter and one receiver). If there are three or more stationary antennas in such a system, I will again kowtow to standard nomenclature and refer to this system as *multistatic*. If one or more of the antennas is moving, I will break with standard practice and call the system *bikinetic* or *multikinetic*.

Bistatic operation is as old as radar itself. In fact, most of the early experiments in radar, the CW interference systems described in chapter 1, were bistatic since the transmitter and receiver were separated. This separation was necessary in the days before pulsed transmissions and transmitter blanking. The Daventry experiment of Watson-Watt is a historical example of bistatic radar operation using a BBC transmitter. Later on in World War II the Germans adopted this idea, making use of the British CH radar transmitters (so-called *noncooperative* transmitters), and their own specially designed receivers, to learn about Allied air traffic in the English Channel. So bistatic operation is not new, but was soon abandoned in favor of *monostatic* (colocated transmitters and receivers) pulsed radars. Many radar systems of the future will be multistatic or multikinetic pulsed-Doppler systems, though this combination of techniques presents formidable difficulties and is, as of now, mostly experimental.

Currently the most widespread example of bikinetic radar operation is found in the guidance systems of *beamrider* missiles. These missiles follow a narrow beam that illuminates the intended target. Typically, the illuminator will be part of the missile launcher, which may be ground based or part of a fighter plane weapons system. The target is usually an enemy plane, which of course will do its utmost to get out of the spotlight, that is, throw off the illuminator beam. However, so long as the target is tracked, the missile need only follow the beam until it bumps into the target. The point is that the radar system receiver—the seeker head within the missile—is separated from the transmitter (the illuminator), and so this is a bikinetic system, though it is very restricted and specialized.

The next generation of bistatic radars will be much more advanced than the simple missile guidance system just outlined. These new systems will be designed for much more general application, such as surveillance over a large volume of airspace. Two groups will be the main beneficiaries of this emerging technology: the meteorologists and the military. First, let me explain why weather radars will be improved by bistatic operation. The envisaged radar systems (which began deployment within the past few years, though they are still experimental devices or prototypes) consist of a single stationary transmitter plus a number of stationary receivers, and so configure a multistatic system, in our terminology. The transmitter illuminates a weather cell, say a thunderstorm, and the echo power is detected by all the receivers in their different locations. Thus the thunderstorm is being seen from many different angles—this provides increased information. In particular, the velocity of any particular packet of air can be inferred accurately. Recall that, for a monostatic system, only the radial component of target velocity (toward or away from the radar) can be determined from the Doppler shift. In a multistatic system, however, there are several Doppler shifts (one for each receiver); from the combined information from all receivers and their known geometrical distribution, the air packet velocities—not just the radial components—can be calculated. Clearly meteorologists will gain greater understanding of what is going on inside a thundercloud, for example, if they know precisely the velocity field within the cloud and can observe this field evolve over time.

The military are interested in bistatic radar operation for this reason—better target velocity information—but for other, tactical, reasons as well. One very important military advantage of bistatic radar operation is that it can be covert. A radar gives away its position to the enemy by the very act of transmitting microwave energy. So the enemy can retaliate by jamming the receivers—sending loud EM noise back toward the source of the radar transmissions. The jammer signal fills up a conventional radar receiver with noise, thus hiding potential targets, as we saw earlier. But bistatic systems are unjammable. The jamming radar can still detect a bistatic transmitter and estimate its direction, but the bistatic receiver is not in the same direction and so the jammer does not know where to send his noise. Another variant of covert multistatic operation borrows from the German World War II idea of noncooperative transmitters. Thousands of microwave transmitters are out there (high-definition television [HDTV] transmitter towers and cell phone stations, to name a few). Imagine a radar receiver tuned in to these transmissions, knowing the precise location of the transmitters. If an object that reflects the microwaves comes within range—such as an airplane—then the echoes from this target will change the nature of the receiver signal. We have all experienced how a plane passing overhead can mess up a TV signal. This idea may sound familiar because the early CW radars described in chapter 1 used the same principle. Future multistatic receivers will utilize transmitters that are designed for an altogether different purpose, and that send out digital signals. The multistatic radar receiver can, in principle, decipher the disturbed transmitter pulses and estimate the target location and speed. The military potential of such a system is significant, since the military component of the system (the receiver) is passive and therefore undetectable. It is also small and portable, since the direction estimation comes from transmitter geometry rather than a highly directional (and thus large) receiver antenna. The transmitters were designed for a different purpose and so they do not reveal the presence of a remote sensor. Imagine a spy going into an enemy city with such a receiver; he may be able to learn about airplane movement or ground traffic simply[2] by tuning into the local TV transmissions. Imagine such a multistatic system in a friendly city: it could provide radar surveillance coverage without interfering with other systems (recall the crowded EM real estate) by swamping them with its own transmissions. In the United States, a multistatic system called Silent Sentry is being developed by Lockheed Martin, based on commercial FM radio transmitters. It will be able to detect and track airborne targets over a large volume of airspace. In Britain there is a similar project that will utilize the digital cellular phone transmitter network.

Another military application (this time, multikinetic, and with a cooperative transmitter) involves a single spaceborne transmitter and several airborne receivers. The satellite transmitter illuminates targets from above, and the receivers triangulate the echo signals to detect and locate the target. The advantage of such a system is that, since the target is illuminated from above, it cannot hide behind hills or over the horizon. Also, stealth aircraft targets will be decidedly unstealthy.[3] There are many other technical advantages to such a system, which, I hasten to add, is still on the drawing board. It is anticipated that the multikinetic geometry will have a large detection range and better tracking accuracy. The adverse influence of transmitter sidelobes is removed, quite literally, since in this case the transmitter is so remote from the receivers.

Bistatic radar operation, whatever the application, has several advantages over monostatic operation:

- The fact that the target is illuminated from several different directions, or that the echoes are received from different directions, means that target classification can be more detailed. .

---

[2]Well, simple for the spy, though not for the radar systems engineers who have to develop the signal processing algorithms, or for the hardware engineers who have to implement them.

[3]Recall that stealth aircraft are designed with radar-reflective features, such as air intakes, on the upper surface so that ground-based radars cannot see them. The fact that the satellite transmitter will illuminate from above, and that multistatic receivers will view the target from several angles, makes it almost impossible to remain stealthy.

- Multistatic and multikinetic systems with many receivers are more sensitive than conventional systems with one receiver, because more of the transmitter power reflects off the target into a receiver.
- Multistatic systems are able to estimate target positions more accurately than are monostatic systems.[4]
- Discounting development costs, multistatic production costs can be small because the receivers can be small.
- Applications such as Silent Sentry do not require scanning antennas with narrow beams. They "stare" at the target: this continuous coverage reduces the possibility of a target breaking lock, that is, of the multistatic radar system losing track.
- Covert systems that utilize commercial UHF and VHF radio frequencies will not suffer much weather degradation, since these frequency bands penetrate precipitation better than microwaves do.
- Multistatic systems that are large with many receivers forming a network will degrade gracefully. That is to say, if one receiver is knocked out, then the others can take up the slack, and so the overall system performance is barely affected. In a military context, as more and more receivers get knocked out, the system will degrade, but not catastrophically. For example, making up some numbers, with 10 of 20 receivers eliminated a multistatic system may still operate with 90 percent of its original effectiveness.

With all these advantages, why is the world not yet knee-deep in bistatic radar systems? Many serious technological challenges exist that have only recently been addressed, if not overcome:

- The most demanding technological requirement is extremely accurate timing, with clocks that are synchronized throughout the system. This problem is particularly acute for certain systems (e.g., pulsed-Doppler bikinetic and multikinetic) where clocks must remain synchronized within a few picoseconds (10^{-12} s) for several hours. For less-demanding applications (noncoherent multistatic networks) synchronization through the GPS satellite system may be sufficient.

---

[4]With one transmitter, it takes three receivers to estimate the position of a target. Position estimates are based on accurate measurement of the time-of-arrival differences, at the three receivers, of a transmitter pulse. If the system consists of more than three receivers, then the extra information provides independent estimates of position; the different estimates can then be averaged to produce a more accurate estimate.

- To estimate target position accurately, all the transmitters and receivers need to see the target. This may be difficult for ground-based systems in hilly terrain.
- The Doppler domain of the processed signal is very complicated because there are several Doppler shifts for each target.
- Multistatic systems are intrinsically more complex, and sorting out the data requires much more number-crunching processor power than is needed for monostatic operation.
- Target cross sections are generally about half as big (i.e., -3 dB) as monostatic RCS because monostatic systems benefit from target corner reflectors.

Finally, let me mention—only mention—the possibility of multistatic SAR. In principle, this can lead to three-dimensional tomography: the radar equivalent of holograms. I might be able to tell you a little more about this intriguing possibility in about 30 years.

Array Developments

Phased arrays are currently being deployed; these are replacing the older type of dish antenna (the parabolic reflectors, for example) in large radar systems. Phased array systems are designed to perform many tasks at once, and this is where the flexibility of phased arrays pays big dividends. We have already encountered linear arrays, consisting of equally spaced antenna elements. In note T4 we saw how such elements, each transmitting a wide-beam signal, combine to form an array with a narrow beam. What I did not explain in chapter 1, but will expand on here, is how such arrays can steer their beams electronically. That is, the beam direction can be changed without physically moving the array. Here is a bad analogy: try staring straight ahead, but concentrate on a moving object by following it with your eyes, while keeping your head still. This is a bad analogy because it suggests that the array elements move physically, even if the array itself is stationary. To improve the analogy, stare straight ahead and concentrate on an object in your peripheral field of vision, without moving your head or eyes.

There is a great advantage to electronic beam steering for tracking radars, for example. Imagine an antimissile system aboard your Navy destroyer (let's make you the captain, again). Incoming enemy missiles try to avoid being intercepted by this system; the incoming missiles throw themselves around the sky very fast, trying to cause your radar to lose track. But your radar has a phased array antenna, and so it can rapidly follow the twists and turns of

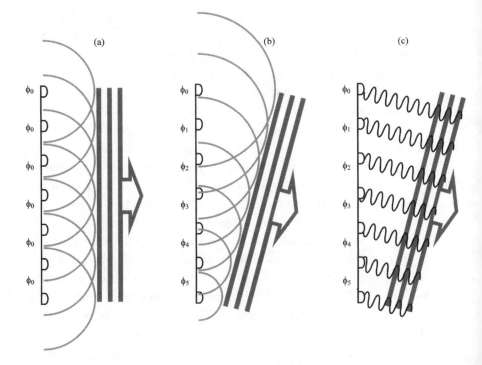

Figure 6.5 *Linear phased array of seven elements, viewed sideways. (a) Each element (black Ds) transmits with the same phase ϕ_0. The individual element wavefronts combine to form a common wavefront (bold gray lines) that define the beam direction. (b) The phase of each element is different—here the differences between adjacent elements are the same. The phase difference determines beam direction. (c) Same beam as (b), but now the individual element waves are shown, instead of the element wavefronts.*

the incoming missile. Earlier mechanically steered antennas had to worry about the inertia of the antenna, which could not be moved instantaneously, but required powerful motors controlled by servo-mechanisms to steer the beam quickly to a desired direction and then hold it in position. Your electronically steered antenna can dispense with all that mechanical paraphernalia and steer the beam instantaneously (well, within a millisecond); no moving parts means no inertia and so no delays.

So how is an array steered electronically? The explanation is simple, and is illustrated in figure 6.5. Basically, the wave emitted by an element of the array is delayed a little (equivalently, it is phase shifted) compared with the wave of an adjacent element. If the relative delays of all the array elements are choreographed correctly, as in figure 6.5, then the result is a beam that points in a direction that depends on the delays. With very fast modern digital electronics, these delays can be controlled accurately, so that the beam is steered in any desired (forward) direction. Figure 6.5 illustrates the method for a linear array of seven elements. Real arrays are two dimensional and can consist of thousands of elements, but the principle is the same.[5]

Phased arrays can do much more than simply speed up beam steering. One array can be configured to form several beams at once—for example, to track several incoming missiles simultaneously—as shown in figure 6.6. The shape of each beam can be controlled to suit a particular purpose. For example, a beam that is narrow in azimuth can be applied to determine the bearing of a missile, whereas another beam that is narrow in elevation fixes the missile elevation. Then a pencil beam, constructed from elements all over the array, tracks the missile or guides one of your beamrider antimissile missiles onto the target. In figure 6.6 I have illustrated multiple beam-forming by dividing the array elements spatially. Likewise, they could be divided in time. That is, each element might form part of one beam and then, on the next pulse, form part of a different beam. Such time interleaving increases the flexibility of phased arrays. Time interleaving is useful if you need many elements to illuminate the target (for example, if the target is at long range), whereas spatial interleaving would be more appropriate if very fast reactions are important (for example, if a number of incoming missiles are very, very close).

Now suppose that your phased array becomes damaged, say by shrapnel from one of those incoming missiles that was intercepted only just in time. We have already seen how distributed arrays degrade gracefully; your phased array will keep on functioning at close to peak performance even if several elements get knocked out of action—the radar beamforming software reconfigures the array elements to compensate for damage.

Confounded by the spirited performance of your phased array guided antimissile system, the nefarious enemy decides to try jamming your radar.

---

[5]One technical complication of electronic beam steering is that the beamwidth changes with look direction. In particular, the beamwidth is narrowest when looking straight ahead, and broadens as the look direction is squinted sideways. This effect arises because, as we saw in note T4, the beamwidth of an array depends on the array length (not the number of elements, but rather the actual physical length). Now, the length of an array seen from an oblique angle is shorter than the length when seen straight on, and so the beam is wider. In practice this effect limits the coverage of an electronically steered array, to within about 45° of straight ahead. So, to attain all-round coverage, it is necessary to have several phased arrays (say forming five faces of a cube), each allocated its own sector of sky. If you see a large boxlike structure prominently displayed on the surface of a warship, it is probably a phased array radar.

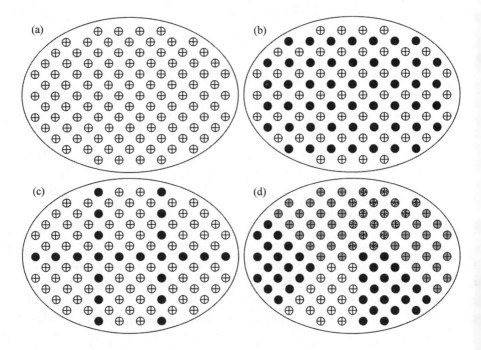

Figure 6.6 A two-dimensional array of 104 elements (real arrays may have thousands of elements), viewed face-on. Different beams are constructed by "spatial interleaving." (a) The entire array is utilized to construct a pencil beam that is narrow in both azimuth and elevation. (b) Two independent thinned arrays with the same beamwidth as before, but with half the power. (c) Nine beams from one antenna. Two independent beams that are narrow in elevation (vertical black lines), one beam that is narrow in azimuth (horizontal black line), and six other beams of intermediate beamwidths. (d) Six independent beams that are fairly narrow in both azimuth and elevation. Interleaving in time is also possible.

Yet again, his evil intentions are stymied by the versatility of your phased array. One consequence of the ability to form multiple beams is this: as well as steering beams in a chosen direction, you can steer *nulls* in a chosen direction. A null is a blind spot that occurs between, for example, a main beam and a sidelobe. (See figure T4.2, for example, to remind yourself about sidelobes and to see some nulls.) Nulls are usually considered as a bad consequence of beamforming, but they are unavoidable. Here, though, nulls are useful. By deliberately steering a null in the direction of the enemy jammer, you are "blinding" your radar to the jammer. The electronic garbage that is being sent toward your radar is not received, even though a missile echo from a nearby location (say separated in angle by only a few degrees) can still be detected. Null steering, like multiple beamforming, can be achieved very rapidly by autonomous fire-control radar processors, and can respond quickly to a rapidly changing battlefield situation.

For airborne applications, large phased arrays are more difficult to achieve because of limited space. One way around this problem in the future will be with so-called *conformal* arrays. These are arrays of radar transmitter/receiver elements that are conformed to the shape of the airframe, rather than forming a linear array or a flat plate. So, the whole side of a fighter plane might constitute a phased array. This is a difficult trick to perform, because the awkward shape of airplanes means that the conformal array will perforce be in three dimensions, and arranging the timing/phase sequencing for beamforming will be hugely complicated. In principle it is possible, but the computational load is too much for even the best of modern radar processors, and so conformal arrays must await developments in computer processing power. Another issue with conformal arrays is this. What do we make the airplane skin out of? It can't be metal, since this would block the microwaves and the airplane radar would be useless. On the other hand, it must be strong, to withstand the buffeting airflow that stresses a fast-moving fighter.[6]

Hi-Res Imaging: The IMAX of Remote Sensing

It is not possible for radar to produce images that are as good as optical images—photographs—as I indicated earlier. We have seen that this is because of the longer operating wavelengths of radar systems. However, the next generation of imaging radars will provide images of higher spatial resolution than current systems (though still not as good as optical), and in addition, will provide other information not present in optical images.

Consider the schematic diagram shown in figure 6.7. Here is an airliner, and a *high-resolution* radar image, assuming square resolution cells. The radar picture is not exactly razorsharp, but it is recognizable as an airliner and is better than an optical image because, let us say, this airliner is 50 km away from the radar, in a cloud, at night. To obtain such small resolution cells (much less than a meter on each side) the radar has to work very hard.

---

[6]This problem already exists for military aircraft with nose-mounted radars (the usual configuration). The nose of the plane consists of a *radome*, which must conform aerodynamically to good engineering design, but which cannot be made from metal. This trade-off—between material strength and EM transparency—is difficult to achieve; mountains of money have been spent to build effective radomes. There are inevitable reflections from the radome back into the radar receiver (very undesirable) and these must be accounted for, in detail, so that the radar processor does not mistake them for target echoes.

Figure 6.7 *Schematic illustration of an airliner and of its hi-res radar image. Here I have assumed that the radar is illuminating the plane from top right. In the image, shade of gray is proportional to echo power. The four checkerboard patterns indicate resolution cells with a Doppler spectrum that is different from the others because of turbofan engine reflections.*

We have seen how high range resolution can be obtained by transmitting linear FM chirp waveforms. Well, *very* high range resolution can be obtained by transmitting *very* wide bandwidth chirps. This doesn't come cheap. The chirp may have to be spread over several pulses—it won't fit into a single pulse. This complicates image formation because the target will move between pulses. Also, very large bandwidths require broadband receivers. Sharp angular resolution is achieved, as we have seen, by synthetic apertures; again there is the problem of how to form a SAR image when the target is moving. Without sophisticated processing, the result would be like the blurred photo of a moving object.

Assuming that all these challenges have been met—and they are being met even as I write—then we can form a coarse radar picture of a moving target, as indicated in figure 6.7. However, hi-res radar images are better than this grainy picture suggests, because of the Doppler dimension of coherent radars images. The airliner is moving at, say, 500 km hr^{-1} and so each echo from each resolution cell will be Doppler shifted from the transmitted frequency to the same degree. However, a few resolution cells will have an altogether different Doppler structure, as indicated in figure 6.7. This difference arises for those cells corresponding to the airliner engine locations, because the hi-res radar can see inside the engine intakes (if the aircraft aspect is obliging) and pick up the rotating fans. Back in the radar signal processor, the characteristic turbofan engine spectrum is recognized and the resolution cells that possess this spectrum are identified as turbofan engine locations. Other moving parts, such as rotating antennas, cooling fans, and helicopter rotors, can similarly be identified via their Doppler spectra. So, we are now developing the capability to form hi-res radar images that reveal more information about the target than its size and shape; we are beginning to be able to discern something about its internal structure. This is the territory of cetacean and microchiropteran echolocation signal processing. From information about target size, shape, speed, and internal structure we are only one step away from target classification and only a few steps away from target recognition; when we get there we will be able to hold our heads up among the dolphins and bats.

Hi-res imaging goes hand in hand with clutter reduction, because image quality depends on signal-to-interference ratio. There is no point in developing images with a resolution of 30 cm at a range of 50 km if the images we obtain are buried beneath clutter or noise. In chapter 3 we explored two techniques that assist with digging a target signal out from beneath strong

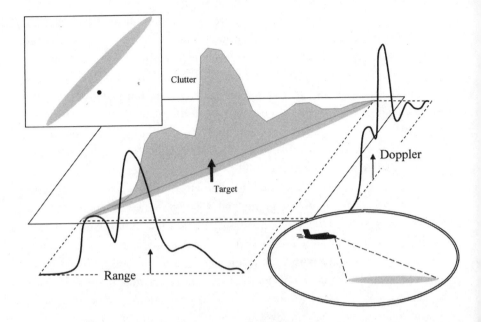

Figure 6.8   *Two-dimensional clutter (gray) and target (black arrow). This type of clutter is common, in particular, for airborne radar looking at the ground (inset). The target and clutter are in fact distinct (plan view of clutter map, at top) but the time-domain projection and the frequency-domain projection cannot separate them. Space-time adaptive processing (STAP) can, by looking at the plan view.*

clutter and noise. We can boost the signal at the expense of noise by integration, or by correlation. These techniques are said to operate in the time domain, since they work with data that is time ordered. On the other hand, we have seen that target and clutter echoes can also be separated in the frequency domain, based on the different Doppler frequency spectra of clutter and a moving target. But sometimes, working in one or the other of these domains is just not good enough to eliminate the clutter; we get rid of some clutter but are still left with fuzzy target images. Cue STAP (for Space-Time Adaptive Processing), a technique that has blossomed in the past few years, and today constitutes one of the most intensely investigated fields of radar signal processing.

I can convey to you the idea behind STAP with a simple picture, if I take a few lines to set the scene. Imagine a clutter source that is obscuring a target signal. In figure 6.8 the clutter source is the ground or ocean. The radar is housed in a plane; perhaps this aircraft is a ground-attack fighter looking for enemy tanks, or perhaps it is an air-sea rescue airplane looking for survivors of a shipwreck. Whichever, the aircraft radar illuminates the surface and receives clutter echoes from it. These echoes are Doppler shifted because of the relative motion of airplane and surface, but by different amounts. You can see from the inset in figure 6.8 that the radial component of speed will be greater from elements of the illuminated surface (the *radar footprint*) that are furthest away from the plane, than from footprint elements that are nearest the plane. In short, the clutter speed depends on range or, equivalently, on elevation angle. Echoes from different ranges/ angles arrive with different time delays and so we can refer to this axis equivalently as the time or range or angle domain. In figure 6.8 I depict a clutter map (of clutter power plotted on a two-dimensional map of range/ angle/time versus Doppler frequency). Note how the clutter is not uniformly distributed over time-frequency space—this is a quite common occurrence. The target exists at a particular location and has a fixed speed relative to the airborne radar, and so occupies a single point in time-frequency space. Having set this scene I can now describe why time-domain processing alone (or frequency-domain processing alone) can be unsatisfactory, whereas STAP processing performs much better.

From figure 6.8 we see what happens if we project the clutter onto the range/angle/time axis. The clutter dominates the target. If we project onto the frequency axis (i.e., if we work with the target and clutter spectrum) then we still have our work cut out to detect the target, since it is still pretty much inseparable from the clutter. However, we know from the clutter map that target and clutter *are* separable, and so in principle we can suppress the clutter without also squashing the target. This is what STAP achieves. STAP processes the echo data in both time and frequency domains; for example, it will consider the frequency components (of an echo signal) drawn from different times, and so can see the plan-view clutter map shown in figure 6.8. This processing is difficult, and only in the past decade or so have we come to grips with it, but now it is possible for us to distinguish between the kind of clutter viewed in figure 6.8 and the target shown. Because STAP can see the plan-view picture, it can appropriately filter the data to remove echo power that originates in the clutter region, leaving the target standing out clearly.

ROVs

Remotely Operated Vehicles take humans out of the remote sensing loop. Or rather they take humans out of remote sensing data gathering and out of danger. Be it mine detection, bomb disposal, or search and rescue in hostile environments (burning buildings, bomb disposal sites, radiation spills . . .), these machines provide us with sensory information at a safe distance. However, they are not the subject that I wish to discuss here because, though technically they count as remote sensing devices, these ROVs do not contribute much to the development of remote sensing processing. A camera on a motorized trolley is a remote sensor, but it is still just a camera.

The most widespread use of ROVs in the twenty-first century will be military UAVs. Unmanned Air Vehicles will provide our armed forces with images of enemy installations and deployment (spies in the skies), and will deliver munitions while operating autonomously over a battlespace[7]: they will be robots of war. These are the vehicles that I wish to say something about, because the probable future development of UAVs will likely require significant strides forward in our remote sensing capabilities.

The Predator UAV (fig. 6.9) started out as a technology demonstrator, and was tested in several roles by the U.S. military during the 1990s: Afghanistan, Bosnia and Kosovo, Iraq and in the Yemen. The pilots who fly predators remotely say that it is like flying a plane while looking through a straw, which gives a sense of the limited visibility afforded by onboard cameras, compared with our own remarkably wide-angle peripheral vision. However, a data link of several hundred kilometers separates a pilot from the danger zone (the link may be via satellite, which would avoid some line-of-sight eclipsing problems in mountainous regions such as Afghanistan). This pilot-safety factor is the main advantage of the Predator and similar ROVs. In today's world, the lives of pilots are valued more highly than they were in decades past. Perhaps they always were valued highly, but only now we are developing the technology to remove the pilots from danger. The Predator has been used for remote sensing ("situational awareness" in modern military speak); it can carry a color movie camera housed in the nose, an IR camera, and a SAR imaging radar. It can stay airborne for 24 hours, at altitudes up to 50,000 feet, and with its suite of sensors is capable of providing images at any

---

[7]Nowadays we tend to refer not to a *battlefield*, the traditional two-dimensional designation of a war zone, but rather to a *battlespace*. This is more appropriate because it includes the air (and perhaps space) above a battlefield: it is a volume, rather than an area. The term reflects the new direction that warfare has taken during the past century.

Figure 6.9 (a) The Predator UAV (unmanned air vehicle), an advanced concepts technology demonstrator from the 1990s, is now deployed widely. It can be operated remotely as a missile launcher or as an eye-in-the-sky. USN official photo. (b) The Fire Scout is one of the first vertical take-off autonomous UAVs. USN official photo by Kurt M. Lengfield.

time of day or night and in any weather (so long as it can fly—bad weather is the main limitation on current ROV performance). The Predator is also capable of operating in a hunter-killer role, with two remotely operated Hellfire air-to-ground missiles, which are guided to their targets by an IR illuminator. The hunter-killer configuration provides "precision targeting support" (modern military speak again).[8]

The Predator is controlled remotely by a pilot and two sensor operators. This ROV technology is clearly an intermediate step between piloted airplanes and autonomous robots. Indeed, the Predator can perform simple surveillance operations autonomously. Much of the current UAV development is devoted to attaining full UAV autonomy. The Fire Scout UAV shown in figure 6.9 has recently demonstrated autonomous take off and landing from a ship, and autonomous fire control of missiles. Autonomy (the ability of machines to make decisions without human intervention) is the way of the future, but clearly it involves significant advances in computer control. Quite apart from automatic piloting, which is a well-developed field, the autonomous UAV must make use of its sensors to assess the tactical situation and make decisions that optimize its effectiveness. For example, it needs to know how to get from A to B in the most efficient way, how to schedule tasks, and, perhaps most demanding of all, how to cooperate with friendly forces such as other UAVs. Autonomous UAVs will extend the military advantages of ROVs; not only will pilots lives be spared, but the relatively inexpensive[9] UAVs will be deployed in much larger numbers. The tactical advantages are enormous. One obvious example: inaccessible regions (such as the extensive mountain ranges of Afghanistan) will not be able to hide enemies. Numerous remote sensing autonomous UAVs will be able to cover large areas 24 hours a day, providing tactical information and flexible military options without endangering (friendly) lives. Future military interventions in hostile territory may be more politically acceptable at home if fewer (friendly) lives are lost.

---

[8]The CIA reportedly makes use of Predators and other unmanned air vehicles (UAVs) for assassinations. The advantages of high-flying UAVs for covert surveillance are clear, so we can expect the intelligence community to adopt wholeheartedly the new remotely operated vehicle (ROV) technology.

[9]UAVs are inexpensive compared with piloted airplanes. UAVs do not need to carry the extra weight of pilots, or the extensive and expensive life-support systems (such as ejector seats, which are very expensive because they must be fail-safe) that look after them. There are knock-on consequences: UAVs are relatively disposable, so that, for example, a lighter airframe is acceptable. With no pilot or pilot support systems and with a lighter, smaller frame the UAV requires a smaller engine and less fuel.

If ROVs such as the Predator represent the current generation, and autonomous UAVs spawned from the Fire Scout and similar development projects represent the next generation, then the next-but-one generation of UAVs will surely be the Micro-(unmanned) Air Vehicles or MAVs. Smaller than 15 cm (6 inches), these are effectively disposable flying cola cans.[10] These devices are already being considered seriously for the future by DARPA.[11] Imagine swarms of these aggressive little beasts flying into enemy formations, using their remote sensors to scout inside buildings, attach themselves to tank tracks and break them, clog enemy plane air inlets, jam gun barrels, attach themselves to buildings, and then either explode or listen to conversations inside. Less aggressively, and less ambitiously, MAVs may distribute themselves throughout a battlespace and, by forming a sensor network, provide detailed information from every corner. We have seen that the complexity of multikinetic sensor networks requires much computer processing power—well, a swarm of networked MAVs will need a further exponential growth of number-crunching capability.

In the latter stages of my career as a radar systems analyst I was involved peripherally with a more modest cola-can project of networked remote sensors. The idea was to distribute randomly over a battlefield, perhaps by airplane, large numbers of cola-can-sized sensors. These sensors would not be mobile, and so would be really inexpensive—thousands might be deployed. The capability of each sensor may be modest: a short-range radar or a passive sonar or IR sensor with limited resolution and range. However, the shear numbers of sensors would yield significant information. Each sends information via data link to a remote central processor that integrates the barrage of data to form a coherent picture. Say that one of the cola cans is capable of nothing more than sending out an electronic beep whenever a large object (such as a tank or a soldier) moves within five meters of it. The capability of each can is modest, but the network exhibits *emergent behavior* and is much more capable than the sum of its parts. By correlating all the beeps from all the cola cans, and knowing the locations of each, the central processor can reconstruct ground target movement of entire forma-

---

[10] You will appreciate that here I am asking you to imagine not only flights of cola cans but also flights of fancy. The idea of cola-can-sized UAVs is futuristic (how *do* bats get so much remote sensing capability into such a small airframe?), in other words, beyond our current abilities. Nevertheless, elements of the imagined sensor network are within our current reach, and so I can plausibly speculate on the nature of mid-twenty-first century battlespace remote sensing.

[11] Defense Advisory Research Project Agency—the central organization of the U.S. Department of Defense advanced projects research and development.

tions. Clearing a battlefield of all these machines, to deny such information about forces deployment, would be a major headache.

Am I being fanciful in imagining myriads of little remote sensors scattered like garbage over the ground, or flying like a swarm of locusts overhead? I think not; the technology—both hardware and signal processing software—is presently conceivable and is thus a reasonable bet for the future. Cola cans will never seem the same again.

Sensor Fusion

One of the buzz phrases of the 1990s, *sensor fusion*, awaited sophisticated algorithms and processing techniques (such as neural networks and fuzzy logic) before it could even begin to work. Sensor fusion is the blending of information from different types of sensors to form a unified interpretation of the environment. A simple example will illustrate what I mean, and will also show the advantages of sensor fusion along with the difficulties of implementing it.

Most domestic burglar alarms today are single sensor; typically they sense movement in a room by Doppler processing an IR signal. IR is better than visible light for this purpose for a number of reasons, but it is not perfect.[12] The next generation of domestic alarms may be improved by incorporating passive acoustical information and correlating it with the IR echo signals. It is not difficult to see that this would, if properly implemented, make a better security system. The number of false alarms would be reduced, since a target must simultaneously produce a false alarm in the signal processing chain of both sensors. For example, a moth flying past the IR sensor might exceed the IR echo and velocity threshold, but it would not be making enough noise, or the right type of noise, to trigger an alarm. A sound from outside may exceed the acoustical threshold, but will not cause an alarm unless it is correlated with a movement within the room. Thus an alarm based on the fusion of IR and acoustical signals would be more robust and more difficult to deceive. Put simply: more sensor types mean more information for the signal processor to work with and so, given the right processing, a more effective system.

---

[12]Glass does not transmit IR and so objects that are moving outside the room window will not cause the alarm to sound (an alarm based on visible light might have this problem). IR sensors are not perfect because the sensor can be blocked, and because it may not register very slow movement. Recall from chapter 2 the trade-off between detection and false alarms. If a threshold speed and echo power is set too high, then slow-moving or small targets (in this case, burglars) may not be detected, but if set too low then other sources such as house plants or spiders may set off a false alarm.

The difficulty with sensor fusion lies in the phrase "given the right processing." Our burglar alarm may be made to work with IR and acoustical sensors operating independently, each with its own threshold. Both thresholds must be exceeded to cause an alarm. This approach is not sensor fusion, however, but sensor multiplication. The main benefit of sensor fusion comes with the mingling of sensory information, and this mingling is very difficult to do. Consider again the burglar alarm. A window being forced open may exceed the acoustical threshold for an alarm, but the alarm should only occur if this sound is accompanied by the window opening—detected by the IR sensor. Otherwise, somebody outside who innocently taps on the window may set off the alarm. The problem is to associate, in the signal processor, the acoustical signal of the window sound with the IR signal of the window movement. For humans this is easy—we do it all the time, and very well.[13] Perhaps because we are so good at it, we might find it hard to understand what the problem is. You might say: "The sound of the window forced open occurs at the same time as the window movement, so clearly this is a break-in and so the alarm should be tripped." But there is much more to it than this: other extraneous noises or movements may also occur simultaneously, and so we must use more sophisticated methods of discrimination than simple timing. Humans must somehow be able to filter out the extraneous signals subconsciously and effectively—the problem is teaching this skill to our signal processors. Sensor fusion signal processing involves learning about what is important and what isn't. In the case of the burglar alarm, for example, the idea of "window" must be given an acoustical and an IR meaning. We must teach our remote sensor signal processors to be more like us, or like bats, so that they can perceive distinct objects in their environment and associate the sensory information (IR or sound) with each object. This is the distinction between sensor fusion and what I have called sensor multiplication.

Sensor multiplication (IR plus acoustics) is familiar and obvious. Rescuers use IR sensors and microphones to help them detect bodies in rubble in the aftermath of an earthquake. The acoustical information and the IR information are synthesized by the rescuer—for example, he hears a sound in the same location as he sees an IR image. Medical imaging ben-

---

[13]Think how much sensor fusion goes on in the brain of a National Football League wide receiver, struggling to catch a football while evading tacklers: wide-angle vision, directional sound, touch.

efits from a multiplication of sensor types: X-ray, ultrasound, MRI, etc. However, in medical imaging and in many other applications much better results can be obtained by true sensor fusion. Terrain navigation is one such application. The sensory information from a car wheel, measuring speed, may be fused with a database (a street map) to yield, with appropriate processing, the location of your car in a city. A cruise missile may use a database map to correct or confirm the course it is following, based on information it receives from both its sensors and its inertial navigation system. A UAV or a submersible may follow a course based on altimeter information and a topographical database. Another application is automatic target recognition. A criminal or terrorist in a crowd, or tanks beneath foliage, may be difficult to identify reliably with only one sensor type. Sensor fusion will provide more robust recognition algorithms.[14]

Consider now the perennial radar task of detecting airborne intruders. A surveillance radar may be ably assisted by satellite optical or IR imaging. The coarse-grained but wide-ranging information provided by radar may be enhanced with finer details obtained by opportunistic visible or IR observations. Clouds may obscure a satellite viewing intruder aircraft, but occasional gaps in the clouds may permit brief detailed observation, with the satellite cameras being directed to the right place at the right time by the radar sensor data. This use of multiple sensors is not really sensor fusion or multiplication, but rather sensor adoption. Our remote sensors of the future will use whatever sensor they possess that will provide the best information under the circumstances of the moment. This intelligent exploitation of circumstances is another example of high-powered processing applied to remote sensing.

Here are two other, quite different, examples. First, SAR images may be exploited to learn about the height of objects that are within the SAR scene,

---

[14]Another example: humans are hardwired to recognize faces. We are superbly good at it. We have evolved many facial muscles to express emotions that are communicated visually to other people. We recognize a face that we have not seen for 30 years, even though it may physically have changed significantly (by growing a beard, for example). Yet we somehow know that face. (We also know that voice, and sensor fusion helps us identify long-lost friends or enemies—a skill that goes way beyond prodigious feats of memory.) It is notoriously difficult to program signal processors to recognize faces, though much academic effort has gone into this field. Progress is being made, spurred perhaps by a perceived need for greater security. When our efforts are ultimately crowned with success, as they will be within a decade or two, those vital elements of sensor fusion processing—neural networks and fuzzy logic—will form an essential component of whatever facial recognition processing algorithms prove successful.

by observation of the object shadow length. Recall that SAR images are formed typically at a grazing angle of a few degrees, so that tall objects cast long shadows. Knowing that a target casts a shadow of particular shape can lead to improved target detection. Second, SAR images of moving objects are often distorted, like a Dali painting, because the Doppler processing that is required for synthetic aperture formation gets confused by the Doppler shift of moving objects within the scene. A well-known example is moving trains. In SAR images the train often appears separate from the tracks, in a field alongside, say, because the train is moving relative to the rest of the scene. Clever processing can put the train back on the tracks. These are not really true examples of sensor fusion, but are similar in spirit: the image processor is making use of extra information not contained within the SAR data. In the second case, we know that trains must appear on tracks, and so images that look like trains, but appearing in cornfields and running parallel to the tracks, should be put back on the rails. Similarly, a tank that appears in the middle of a lake is probably moving along a shore road; cars that move parallel to highways, but appear to be sitting on the tops of buildings, should really be back on the road.

Sensor fusion is, at bottom, a sensible strategy of including all possible types of information that can tell us something about the target or scene that we are imaging. Once achieved—and this will be an ongoing process for many decades—then remote pilots will no longer feel like they are looking through straws. Not, you understand, because the remote pilots will be presented with a wider view, but most likely because they will be dispensed with altogether. Sensor fusion and autonomous UAV operation go hand in hand.

BIBLIOGRAPHY

Specialist applications produce specialist literature, not suitable for the general reader. There are a few educational Web sites that are worth consulting, however.

For weather radar principles, the University of Illinois Urbana-Champaign Web site http://ww2010.atmos.uiuc.edu/(Gh)/home.rxml is very accessible. There are many weather-forecasting sites on the Web with Doppler radar imagery.

A brief history of air-traffic control (ATC) radar and a PPI image is found at the IEEE Web site www.ieee-virtual-museum.org/collection/tech.php?taid=&id=2345899&lid=1. A much more detailed, though quixotic, account of ATC radar is provided by Wikipedia at http://en.wikipedia.org/wiki/Category:Air_traffic_control.

A collection of echocardiology Web sites, some showing echocardiogram animations, is provided at www.rtstudents.com/echocardiology/.

Similarly, advanced techniques tend to be represented in the literature by advanced textbooks, with very little written for the nonexpert (hence, this book). The following STAP reference is available for the dedicated student:

Klemm, Richard, ed. *Applications of Space-Time Adaptive Processing.* Institute of Electrical Engineers, London, 2004.

The literature for sensor fusion and other advanced techniques is either too technical or too superficial to be of interest to many readers. For those who are interested in performance specifications several Web sites by manufacturers and end-users (i.e., the U.S. Army and Navy) offer facts and claimed performances of several UAVs.

7............ Final Thoughts

Please understand that, despite the twenty-three technical notes and the intentionally broad-sweep, scatter-gun approach that I have adopted for this book, I have *only scratched the surface* of remote sensing techniques. Also, there are alternative versions of every single technical subject discussed in this book, from antennas, beamforming, CFAR, and Doppler processing right through to the end of the remote sensing alphabet. My aim has been to provide you with an appreciation of the processing that underpins remote sensing—whatever the application may be—at a technical level that is deeper than my "Math-Lite" approach may imply.[1] So much for the technical challenges of remote sensing. Here I would like to indulge myself a little—vent a spleen, you might say, once you've finished reading this short chapter. As promised in "Hearing the Picture," I will say a few words about the organizational aspects of the remote sensing industry, and about how such (in)efficient (dis)organization helps (hinders) significantly the practical implementation of working remote sensing systems.

---

[1] If, because of the absence of math, you doubt the technical content of this book, please glance at the Glossary. I have succeeded if most of the remote sensing words now mean something to you. ("I've always wondered what that meant!")

Managerial Entropy

The small passenger plane was a noisy twin-propeller Fokker, with overhead wings and a very narrow fuselage. We were on our way from Edinburgh, Scotland, across the Irish Sea to Belfast, for an early morning project meeting with a partner company. I pointed out to Stuart, my supervisor, the beautiful scenery of the Scottish Isles, which at that moment formed a ring of hills around us, snow-capped and lit up by the rising sun. Stuart cast a bleary eye out of the window at this magical scenery, and remarked that the snow reminded him of the scum left behind in a bathtub.

Now Stuart is one of two radar engineers that I have had the pleasure of working with whom I could describe as genuinely brilliant, a multitalented engineer and an essential member of any design team. His one professional failing (I do not include his lugubrious sense of humor) as he would readily admit—and as anyone who has worked under him would more readily admit—is that his people-management skills were AWOL. He had been promoted through technical ability to a position of management, for which he had little ability, and which diverted him from the purely technical role at which he excelled.

Fast forward ten years to a different company—a very large UK-based multinational aerospace corporation. By the 1980s many radar projects were becoming behemoths and were far too large and expensive for a single company, even a multinational one, to tackle alone. We had entered the era of consortia, with different companies (often from different countries) getting together via contractual agreements of Byzantine complexity to design and build very large and very complex radar systems. In my case, our corporation formed part of an agreement to build the next generation of European fighter aircraft—the Eurofighter—and we in the radar division were in partnership with other radar divisions from other companies to build the advanced multimode pulse-Doppler radar system for this beast. The end result for both airplane and radar system will be very impressive but—oh my—what a way to run a business.

There are difficulties enough with consortia, since the various partners have somewhat different agendas, and because the legal entanglements (subcontracts with progress meetings and payment milestones) necessarily involve extra layers of bureaucracy. Progress is always slow. Engineers are weighed down by paperwork and paper pushers, and feel like they are wading through molasses in a snowstorm. Tempers get frayed, and fingers of blame get pointed. Add to this fraught setup the complexity of the large radar system that is being developed, with its design intricacies and real technological challenges of implementation, and you can see that (almost inevitably) a difficult labor precedes the birth. In our case, this situation was made worse by the multi-national nature of the consortium, because different partners wanted different versions of the final product to meet different national needs.

Everything I said in the preceding paragraph no doubt applies very widely in industry, and much of it is unavoidable, but it seems to many of us foot soldiers that the generals and their staff officers are exacerbating matters. I do not refer here to the Stuarts of the corporate world, since their technical aptitude more than compensates for their management foibles, but rather to the myriad of less talented middle managers. These very ordinary people are usually promoted from the ranks as a reward for ambition or for hard work or long service in a technical field, rather than because it is thought that they would do a good job as manager. I hypothesize that this group of people clogs the organizational arteries, and is largely responsible for much of the inertia that bedevils large engineering corporations. Let me take as a random example the fictitious Mr. Klutz. He will have performed a competent though undistinguished role as an engineer for several years. Mr. K, an ambitious man (I like to think of him also as being pompous and opinionated), has been rewarded with a junior management position. Freed from direct responsibility for technical work, he is able to devote more of his time to self-promotion. The result is that his position becomes bombproof: he can readily outmaneuver engineers when things go wrong, and point the finger more effectively, since he is not devoting his time to solving technical issues, but instead has more time to think about self-preservation. As adept at taking credit for others' work, he becomes comfortably settled.

Most managers are not like Mr. K, I hastily add, but it does not take many like him in a large organization to sour working relationships and departmental effectiveness. All my experience in industry has been with UK-based companies, and there are Mr. Ks in each of them. I have reason to believe that they exist elsewhere in Europe: Monsieur[2] K, Herr K, Signor K. U.S.

---

[2]One of the delights of international radar conferences is seeing how they reflect the national culture of the hosts. I recall a wonderful conference in Brest, France. It was not well organized (for example, the conference Proceedings were delivered late and arrived without page numbers, and there was no organized transport) but the lunches were superb. Providing wine with every lunch meant that many of the afternoons were wasted. The conference dinner was lavish, though one Monsieur K became drunk and turned on a couple of us. I received a tirade because my company had recently, through some Machiavellian boardroom machinations, angered his company. A U.S. radar consultant, sitting next to Monsieur K, then got it because President Clinton was at that time being hauled over the coals by the U.S. media over his affair with an aide. C'est la vie. U.S. conferences are huge and very well organized, but with insufficient wine. German conferences have beer, wine, and organization; British ones have bad food.

companies have a more focused work ethic, in my opinion, and yet they also have problems with large-scale projects. Perhaps because of the Mr. Ks, or perhaps for entirely independent reasons, there appears to be a limit to the complexity of large engineering projects that corporations can complete successfully.

In the field of radar, this upper limit has been revealed in an amusing way by a couple of gurus. Eli Brookner has written a book that deals with many, varied technical issues of radar design, but devotes one chapter to a (frankly hilarious) account of developmental disasters. He shows photographs of huge radar dishes, part of a large and complex project, which catch fire because someone, somewhere made a booboo. Cue for photo of the rebuilt radar dish, with large water tower sited nearby — just in case. The same chapter contains a tongue-in-cheek account of how large projects overrun in time and budget, and so are subjected to cutbacks: less ambitious versions of the original project are put in place. Yet — to no one's amazement — the resultant savings are not proportional to the cutbacks: you get 50 percent of the product for, say, 70 percent of the cost. By extrapolation, you might get 0 percent of the original dream product for 40 percent of the cost. Another tongue-in-cheek extrapolation shows that, if large military aircraft project costs continue to rise as they have in the past, then the next generation will have to spend the entire U.S. budget to build a single fighter plane.

I mentioned David Barton in an earlier chapter. In his radar course he spent a little time opining to us about the woes of large-scale radar projects. He put it bluntly: *no large and complex radar system works as well as it was designed to do*. His idea is interesting; it resonates with anyone who has spent time on such projects.[3] He claims that the complexities are too numerous and too great for any corporation or (in particular) any government to manage. This seems to be suggesting that *there is a sociological, rather*

---

[3]Barton gave an example from personal experience. The prototype of a very large ground-based radar surveillance system appeared to be producing some signals that could not be explained. They were not target echoes, and they appeared only intermittently. Algorithms were checked — nothing wrong there. Wiring and other hardware were checked — nothing wrong there either; the hardware was implementing the algorithms as designed. Calculations were recalculated and further trials made. The fault appeared daily but, someone noticed, only once a day: when the moon rose above the horizon, passing through the radar transmitter beam. The strange signals were echoes from the moon. (The signal time delay was so long that it was difficult to identify the source by its range, since the range greatly exceeded the radar ambiguous range — see notes T2 and T23).

than a technological, limit to the complexity of an engineering system that humans are capable of building. I presented Mr. K as the sociological fly in the organizational ointment, as the barrier to progress. I may be wrong about this; perhaps something else is responsible, but whatever the cause it certainly seems to be true that, in the world of remote sensing, project snafus are proportional to project complexity, which is proportional to project manpower, which is proportional to the number of Mr. Ks.

Progress Report

Our remote sensing capabilities have come a long way over the past 70 years, the preceding gripes notwithstanding (though they will be "withstanding" in the future, if remote sensing systems continue to get bigger). From fuzzy blips appearing on a wobbly CRT tube at the end of the CH radar system we have now advanced to a level of sophistication approaching that of furry flying mammals. This rate of progress is quite astonishing. I would like to think that it has been fueled by our insatiable human curiosity about the environment we live in, and on, and under. More prosaically I recognize that a military imperative drove our original quest for remote sensing, impelled the further development of radar and sonar systems, and continues to propel forward our ability to hear the picture.

So much for the experts who contributed to the development of this large field. What about you? Those of you who read this book as an auxiliary textbook, to gain a university credit, deserve a WMD[4] Bronze Star. If you waded through, and benefited from, the twenty-three technical notes you are to be commended and merit a Silver Star. My Gold Star is awarded to those inquisitive bibliophiles who picked up and read my book simply out of idle curiosity. Mine is an arcane discipline; I have myself benefited from reading accounts of similarly arcane subjects about which I previously knew little (the development of malaria cures, for example, or the history of the Ottoman Empire). I hope that you have gained something from reading my necessarily cursory semitechnical exposition of remote sensing. If so, then perhaps you can return the favor by writing a book about your own area of expertise.

---

[4]WMD here refers to my initials; it does *not* stand for Weapons of Mass Destruction. (Perhaps better remote sensor capabilities would have provided more accurate information about the latter subject.) You may be amused to hear that in the summer of 2005 my wife and I flew from our home in Western Canada for a vacation in New Mexico. My luggage is emblazoned with my initials (and has been since well before the tragedy of 9/11 and the rigors of Homeland Security). It arrived 8 hours after we touched down in Albuquerque—and had been searched.

BIBLIOGRAPHY

Brookner, Eli. *Radar Technology.* Artech House, Dedham, Massachusetts, 1977.

Peter, Laurence J. *The Peter Principle.* Buccaneer Books, New York, 1969. Asserts, with many examples, that in a large organization people tend to be promoted to their level of incompetence.

Chapter 1. Early Days

T1. Waves

The most familiar examples of waves found in nature are those we see moving over a water surface and those we hear as sound traveling through the air. To a physicist or engineer, these waves and most others are described mathematically in the same way. A wave is specified by a *wavelength* λ (or frequency f) and amplitude A, as shown in figure T1.1. For water waves, the amplitude is simply the height of the wave above or below the average water level, and the wavelength is the distance between peaks. So, if in figure T1.1 the horizontal axis represents distance, then the wave shown is a snapshot of a moving water wave, capturing the wave at a particular instant of time. On the other hand, if the horizontal axis represents time, then the wave shown represents the height of a particular water molecule bobbing up and down on the surface. For sound waves, the amplitude is not a height, but rather a pressure. I illustrate this schematically in figure T1.2, where we see an electronic speaker diaphragm on the left oscillating, pushing air to the right. Air molecules become bunched or compressed (dots close together) by the diaphragm moving right, so the air pressure is high, followed closely by regions of rarefied or less-dense air (dots far apart) caused by the diaphragm moving left. A wave of oscillating air pressure, which we call sound, is established as a result of the energy input to the air molecules by the diaphragm. This wave

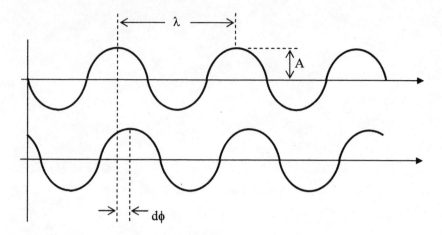

Figure T1.1 *Wave characteristics: amplitude A, wavelength λ, and phase difference dφ. The power carried by any wave (EM, sound, water) increases with amplitude.*

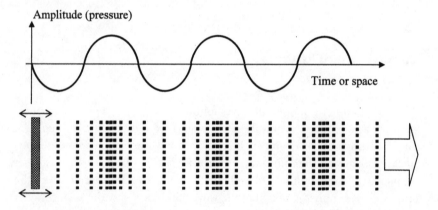

Figure T1.2 *For sound waves the amplitude measures overpressure (pressure above or below average). Here a diaphragm oscillates (left) and each push to the right generates compressions that propagate at a wave velocity that depends on the medium. The wave frequency (and so the wavelength) is determined by the diaphragm oscillation rate.*

moves (to the right, here) with a speed that is characteristic of the medium. So, sound waves in air move at about 330 ms^{-1} (say 1,100 feet per second), and about five times faster through water. Sound waves travel still faster through solids such as steel or rock. Note from figure T1.2 that the pressure oscillations are represented by a wave just like that of figure T1.1, only the amplitude refers to air pressure (above or below the average value) rather than water height (above or below average).

You may well be struck by the following thought: if waves require a medium, such as air or water, to travel through, then how can EM waves travel through space? Space is after all a vacuum, that is, nothing. Good question. Physicists puzzled over this point for most of the nineteenth century until Maxwell shed light on it, so to speak. He showed that, roughly stated, the electric and magnetic components of EM radiation each form the medium for the other. Thus, the magnetic component results from the oscillating electric component (and vice versa), just as water waves are formed by—are—the oscillating water surface. This is why EM waves are the only waves that do not depend for their existence on any external medium.

In addition to wavelength and amplitude, we may talk about the wave *phase* ϕ, meaning the state that the wave is in at a given instant. Wavelength is always a length, specified in meters or centimeters or microns, whereas phase is an angle, usually specified in degrees. The *phase difference* $d\phi$ between two similar waves is shown in figure T1.1. These waves have the same wavelength and amplitude, but you see that one of them is displaced slightly from the other. A phase difference of 0° means that the two waves line up; 180° corresponds to the *crest* (peak, or maximum) of one wave lining up with the *trough* (minimum) of the other; 360° means that the waves again line up, but displaced by one wavelength, so the two waves look the same. Only phase differences usually matter—the absolute phase of a CW transmission is unimportant.

I now come to a hugely, overwhelmingly, megaimportant, world-shattering, epoch-making, common-cold-curing, budget-deficit-reducing property of waves. Sorry for the prolix hyperbole, but I am trying to get your attention. You may have mentally wandered off, a little, during my brief discourse on waves, because the subject may seem to be yawn inducing, or may already be familiar to you. However, I need you to focus on the phenomenon of *wave superposition*, because it is vital to an understanding of radar and sonar systems. (Superposition is also a vital component of quantum physics and is responsible for much of the weirdness exhibited by quantum objects.) In figure T1.3 you see two waves with slightly different wavelengths. Suppose they are water waves, set up by two rocks thrown into a pool. What happens when the two waves meet? Well, they simply add up. The two amplitudes add together to form a new wave that may look very different from the originals. In figure T1.3 we see the result of adding together the two waves with slightly different wavelengths: this *superposition* has a similar wavelength, but with the amplitude *modulated* in a manner that varies regularly with time.—When two tuning forks with slightly different frequencies (and hence different wavelengths) are sounded together, we hear this modulation as a *beat frequency*; this term is used commonly in physics and engineering even when the two waves are not sound waves. Superposition is common to all types of wave:

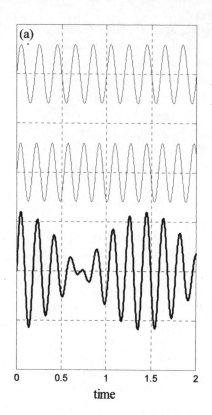
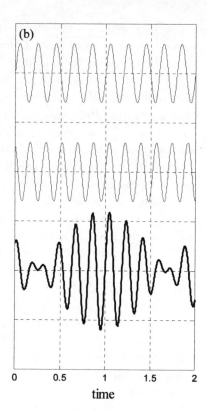

Figure T1.3   *Superposition and wave interference are illustrated when two waves over-lap: they simply add amplitudes (bold lines). The resulting wave can look very different from the originals. (a) Two waves of nearly the same frequency are added to produce a wave that is modulated at the beat frequency. Beat frequency is simply the difference between the two original frequencies. (b) Changing the phase difference between the component waves merely shifts the modulated wave and does not change its form.*

EM as well as water or sound. As I will show you, radar engineers can create transmitter and receiver beams of different shapes by applying wave superposition.

Note also, from figure T1.3, what happens if the phase difference between the waves is changed. We get the same pattern, but it has been shifted in time a little. For CW radars, where the wave is transmitted continuously, this shift is unimportant; what matters is the wave interference pattern caused by superposition of two waves. Recall that for a CW wave-interference radar one wave travels straight from transmitter to receiver while the second wave is reflected off an intervening ship or airplane target before being detected by the receiver. This reflection causes the wavelength to change a little (the "Doppler

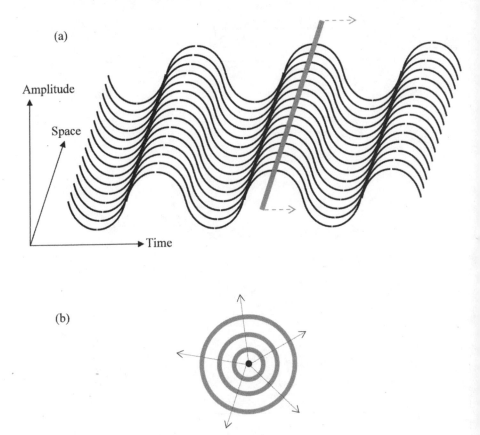

(a)

Amplitude

Space

Time

(b)

Figure T1.4 *(a) A spatially extended wave source, such as a transmitter antenna, can produce a straight wavefront. The wavefront is simply a line of constant phase, and is perpendicular to the direction in which the wave travels. (b) So, for a point source of waves—or for an extended source that is viewed from afar—the wavefront is circular (spherical in three dimensions) since the waves are emitted radially.*

shift," which I will get to later on). The interference pattern is interpreted as indicating the presence of an unseen reflector.

One last fundamental characteristic of waves: you need to know what I mean by a *wavefront*. Suppose I line up a thousand tuning forks, and strike them all at the same time, so that they send out sound waves with the same phase (i.e., zero phase difference), as shown in figure T1.4. The waves are like soldiers marching in step. The wavefront of this line is shown: it is the line formed by joining together the wave crests (or troughs—or any other points in the wave with the same phase). So, the wavefront lies perpendicular to the direction in which the wave moves, and wavefront lines are separated by one wavelength.

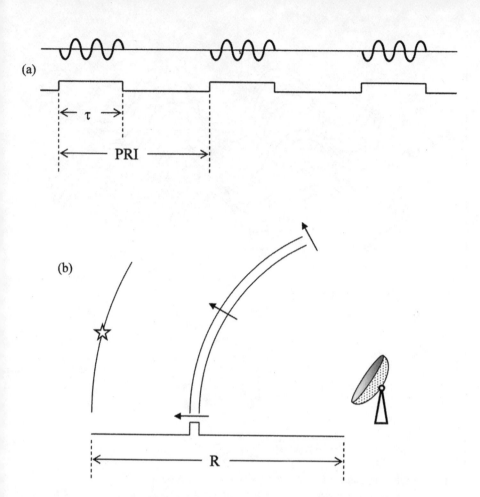

Figure T2.1 *(a) Pulsed wave and the corresponding wave envelope. τ is the pulse duration and PRI is the pulse repetition interval. (b) A single pulse is emitted by radar. The range R to a target (star) is determined by measuring the delay time between pulse emission and pulse reception. The range estimate accuracy is better for short-pulse duration than for long-pulse duration.*

Wavefronts are not always in straight lines. Also in figure T1.4 you see the circular wavefronts formed by a rock dropped into a pool.

T2. Pulsed Waveform, Target Range, Accuracy

We use the word *waveform* to describe the shape of the signal transmitted by a radar or sonar. These waveforms can become very complex, as we will see, with frequency and amplitude both varying in time. Historically, the first pulsed radar waveforms were very simple; they are shown in figure T2.1. A

CW wave of constant, unchanging wavelength (and so with a fixed frequency) is generated; this wave is then switched on and off very rapidly, to produce the type of waveform shown. Instead of drawing the wiggly oscillations, it is convenient to show merely the pulse *envelope*. The pulse duration τ is usually very short.

It is easy to see how pulsed radars can estimate the range of a target. Figure T2.1 shows a radar transmitter that sends out a single pulse of duration τ. This pulse encounters a target at range R, and reflects off it back to the radar. The length of the pulse is $c\tau$, where c is the speed of light, so that a pulse duration of 1 μs corresponds to a pulse length of 300 m. Range is estimated by accurately measuring the time taken for the pulse to radiate out from the transmitter, reflect off the target, and come back to the receiver. In the radar business this travel time is called the *delay time*; let's call it T. So, twice the range is traveled in a time T, or $2R = cT$. So the target range is estimated from the delay time via $R = \frac{1}{2}cT$. For example, if the delay time is measured to be 200 μs then the target range is 30 km.

A potential problem exists here, which real radar systems must take into consideration. Suppose the pulse repetition interval (PRI in fig. T2.1) is t_{PRI}. The target at range $R_1 = \frac{1}{2}c(T + t_{PRI})$ will appear to be at the same location as a target at range R. So will a target at range $R_2 = \frac{1}{2}c(T + 2t_{PRI})$. In general, a target at range R_n will appear to be at range R, because its echo is not from the last pulse transmitted, but from n pulses earlier. The distance $R_{amb} = \frac{1}{2}ct_{PRI}$ is known as the *ambiguous range* of the radar.

The range estimation process is not perfect, however. Even if the delay time is measured accurately, with little error, there is still a limit to the accuracy of our pulsed radar range estimate. This is because of the finite pulse length. From the foregoing you should be able to see that the trailing edge of the pulse returns to the radar receiver later than the leading edge, limiting the accuracy ΔR of our range estimate to half the pulse length: $\Delta R = \frac{1}{2}c\tau$, here 150 m. Short pulses are more accurate than long ones. This will lead to a trade-off in radar design, because shorter pulses have less energy than long ones and so can be detected by the radar receiver only at shorter ranges. So, radar detection range and radar range estimate accuracy work against each other.

T3. Direction Angles, Beamwidth, Beamshape

For now, let us imagine that radar beams are formed in the same way as a searchlight beam, as shown in figure T3.1. A source of light (black circle) is shielded so that it can emit only onto a curved mirror. The light reflects off the surface to form a beam of limited width (shown in the figure by the rays with arrows). The angle of the beam center above the horizontal is the *elevation angle*, denoted by the Greek letter ψ in this book; the beam has

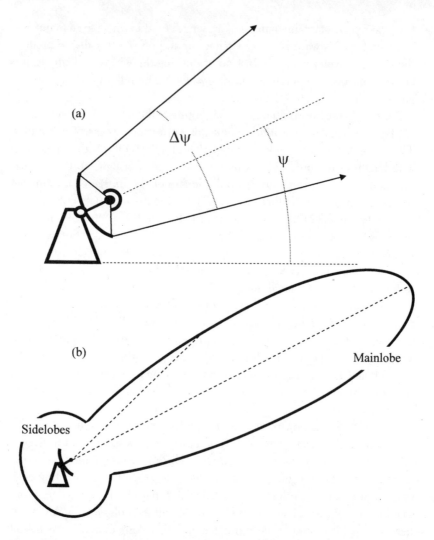

Figure T3.1 (a) A searchlight beam is formed by reflecting a source of light off a curved mirror. The beam direction ψ and the beamwidth Δψ are shown. (b) The beamshape is determined by the mainlobe and the unwanted sidelobes. Beam intensity is represented by the length of the dashed lines, which change with angle.

angular width (*beamwidth* for short) Δψ, as shown. In practice few radars and sonars form beams in this way, but this searchlight analogy is sufficient for the time being to convey the idea of beam direction and beamwidth.

More realistically, but still not quite right, is the *beamshape* shown in figure T3.1. The beam *mainlobe* contains most of the transmitted energy—it is what we see when looking at a searchlight beam—but there is also a lesser amount of energy leaked out sideways and backward. These unwanted leakages in all directions, apart from the mainlobe beam direction, are collectively referred to as *sidelobes*. To radar systems designers, sidelobes are a pain, for reasons that will become clear later on. Figure T3.1 shows a much simplified version of sidelobes; real radar transmitters have more complicated sidelobe structure. The beam intensity is here represented by the distance of the beam outline from the transmitter—longer distance means more intensity or transmitted power in that direction. So sidelobe power is much weaker than mainlobe power. Note one realistic and important consequence: the mainlobe does not have a uniform intensity across it. Unsurprisingly, the beam is brightest right along the mainlobe center and falls away, in a manner than depends on transmitter design, across the beam. So, in the beamshape of figure T3.1 we have a long dashed line representing the angle of maximum beam intensity, and a shorter dashed line representing the angle where the intensity is half as much, since the line is half as long. By convention, we say that the beamwidth is twice the angle between these two lines.

Everything in figure T3.1 refers to elevation, but the same thing applies to the other direction angle, bearing or *azimuth*, denoted θ. Thus the beam outline of figure T3.1 is really three-dimensional, and looks a bit like a baseball bat, or a blimp. For radar beams, however, it is not necessary for the beam cross section to be circular, like a baseball bat. Expressed another way, the elevation beamwidth $\Delta\psi$ and the azimuth beamwidth $\Delta\theta$ may be, and usually are, different.

T4. Beamforming

Transmitters can combine to form an antenna. The waves that are emitted by each component transmitter combine to form beams. The analysis of this process can be mathematically complex, but the effect can be demonstrated simply, by superimposing diagrams. Consider the simple radar transmitter of figure T4.1a. It is a point source (or a cable seen end on) and emits EM waves radially outward; the wavefronts are shown. Let us say that each wavefront marks the position of the crest of a wave. So we have waves radiating outward like water waves diverging from a rock dropped into a pool.

Four of these figures are superimposed, and are shown in figure T4.1b. This corresponds to four transmitters in a line, equally spaced. The spacing is $d = \frac{2}{3}\lambda$. The combined transmissions of these four elements are shown. Because of wave superposition, the beam formed by this *array* of transmitter elements is no longer radial. Note how, perpendicular to the line of transmitter elements, the wavefronts pretty much sit on top of one another. In this region (straight in front of the antenna array, along the antenna *boresight*) component waves simply add up, forming a bigger, more powerful wave. On the other hand, along the line of the array (straight up or down the page, in

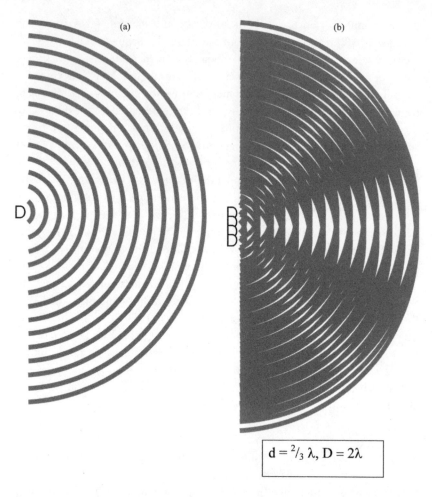

$$d = {}^2/_3\, \lambda,\ D = 2\lambda$$

Figure T4.1 *(a) Transmitter array element and emitted wavefronts. (b) The same picture repeated four times, resulting in a 4-element line array of length 2λ, i.e. two wavelengths, with superimposed wavefronts.*

fig. T4.1) the wavefronts blur out; the crests from different elements occur at different locations, sitting on top of the troughs from other elements so that the waves add up to nothing much—zilch. In between, at about 45° in this case, there is evidence of a weaker addition of component waves, better than zilch but not as powerful as the boresight beam. Between 45° and boresight there is another zilch region.

The high-power boresight region is the mainlobe; the weaker beams out to the side are the sidelobes. I have calculated a realistic beamshape for a linear array and plotted it in figure T4.2: compared with our earlier simplified

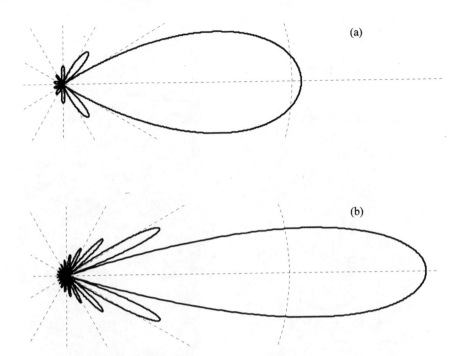

Figure T4.2 (a) Calculated beamshape of a linear array: mainlobe plus sidelobes. (b) Calculated beamshape of a linear array twice as long: narrower mainlobe and more complicated sidelobes.

---

beam, figure T3.1, you see that the sidelobe structure is messier. Now you know why. An antenna that is made up from a linear array of transmitter elements has, because of wave superposition, a mainlobe that is directed perpendicular to the array, along boresight, and sidelobes spread outward at wider angles.

How does the beamshape depend on the array element separation d? In figure T4.3a I have redrawn the array of figure T4.1 but with separation $d = \frac{1}{3}\lambda$ instead of $d = \frac{2}{3}\lambda$. The array length, D, is the same in both cases: $D = 2\lambda$, so that the antenna size equals twice the radiation wavelength. Compare this with figure T4.1 and you will see that the mainlobe looks the same, but the zilch region covers the rest of the semicircle. In other words, the sidelobes are squeezed out of the forward direction. Analysis shows that the critical separation for this to occur is $d = \frac{1}{2}\lambda$. It is desirable to minimize sidelobes, because they waste power by sending radiation in a direction you don't want, and for other reasons that I discuss in chapter 4. So, in practice arrays are constructed from elements that are separated by one-half wavelength, at most.

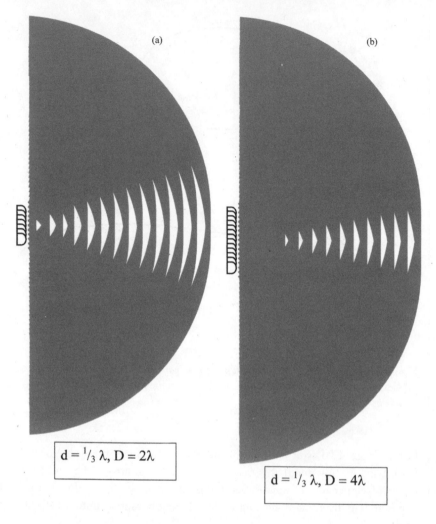

Figure T4.3 (a) Seven-element array of length 2λ, with superimposed wavefronts. (b) Thirteen-element array of length 4λ, with superimposed wavefronts. Note the narrower mainlobe width.

How fat is the mainlobe? Again this is easy to estimate by replicating the basic element wavefronts of figure T4.1. In figure T4.3b I have put together an array of length $D = 4\lambda$, with elements separated by $d = \frac{1}{3}\lambda$. So this array has the same element separation as that of figure T4.3a but is twice as big. Note that the mainlobe region is about half as wide. In fact, the mathematical relationship that tells us the approximate beamwidth of a linear array is

$$\Delta\psi \approx \frac{\lambda}{D}$$

$\Delta\psi$ is the beam angular width (in radians). To confirm this relationship I have again calculated a realistic beamshape and plotted it in figure T4.2. Note that the mainlobe is indeed half as wide (and with more complicated sidelobes). Antenna designers would like the beam to be thin, like a pencil pointing along boresight. This concentrates all the power in one direction (so a target can be seen further away); it also narrows the beamwidth (so the target direction can be estimated more accurately). This is why short wavelengths are favored by radar designers. To obtain a narrow beam at long wavelengths requires a BIG array—hence the large Chain Home towers.

In practice it is not possible to direct all the energy down a thin pencil beam; most of it goes where you want but some just pops out at odd places. It is like a fat lady trying to squeeze into a tight dress: some of the bulges are where she wants them to be, but the rest have to come out somewhere else. There is a great art and science to antenna design: mainlobe beamwidth and sidelobe number and power can be tailored by tinkering with the array element separation. This can become complicated in three dimensions. Our linear array can be extended into a flat two-dimensional surface or a curved surface, and such antennas have three-dimensionalal beamshapes, with the unwanted sidelobes spreading all around.

T5. Angular Resolution

We have seen that *aecuracy* is the precision with which a target can be located. Thus a range accuracy of $\Delta R = 100$ m means that a target range can be estimated to within ± 100 m of its true range. A bearing accuracy of $\Delta\theta = 3°$ means that the azimuth direction of the target is known to within $3°$ of the actual direction. *Resolution* refers to something quite different. Consider two targets—say airplanes—at the same range and altitude. One of them is at a true bearing of $20°$ with respect to a search radar, whereas the other is at $22°$. If the radar azimuth resolution is $\delta\theta = 2°$ or less then the two targets will be seen separately—the targets are said to be *resolved*. If, on the other hand, the radar azimuth resolution is $\delta\theta = 3°$ then the two targets will be detected as a single target; they are not resolved.

I will use the Greek letter Δ (capital delta) to denote accuracy, so that Δx is the accuracy of measuring x, where x might be azimuth or elevation angles, range, or target speed. To emphasize that resolution is not the same I will denote it by a different symbol δ (lower case delta) so δx is the difference in x required before two targets can be resolved. Contrast this with the notation dx that I use, in general, to denote a small difference in x (for example, phase difference $d\phi$). Keep this notation in mind (I will not impose many more such mathematical hieroglyphs on you) and the following technical waters will be smooth sailing.

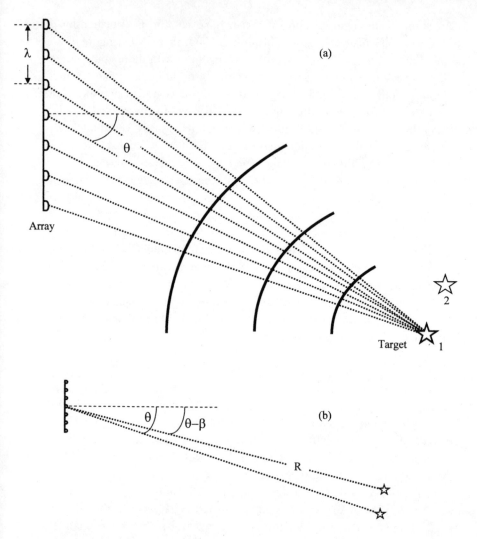

Figure T5.1 *(a) Rays of reflected radiation from one target to the elements of a receiver array antenna. Note the different ray lengths. (b) Two targets, offset from antenna boresight. If the angle β is smaller than the antenna resolution, the targets will not be resolved.*

Without some clever signal processing (pulse compression, chapter 3), the radar range resolution is half the pulse length, as we have seen. The angular resolution is just the mainlobe beamwidth. (Again, this can be improved by clever signal processing, as discussed in chapter 3.) I can illustrate the angular resolution for a linear array antenna as follows. In figure T5.1 two targets are illuminated by a linear array antenna. This time I will consider a receiver

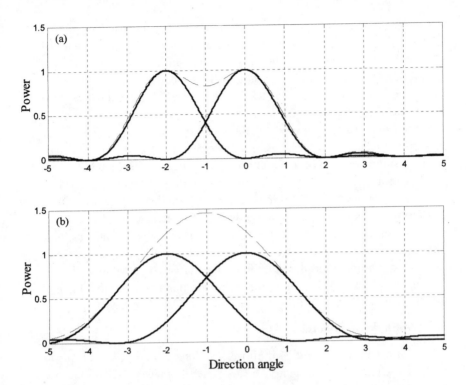

Figure T5.2 (a) Power versus direction angle, relative to boresight, for a linear array of 57 elements looking at two targets separated by 2°. The sum of the two target contributions (dashed line) has two peaks, and so the targets are resolved. (b) The same, but for an array of 35 elements. Now the sum yields only one broad peak, so the antenna detects one large target, not two targets.

antenna (instead of a transmitter) and I show the rays that are reflected off one of the targets, back to each of the array elements. Note that these rays are of different lengths, so that the phase of the wave is different at each element. (Also shown are three of the reflected wavefronts.) The first target is at a bearing angle of θ with respect to antenna boresight, whereas the second target is at $\theta-\beta$ (also shown in fig. T5.1). If β exceeds the azimuth resolution $\delta\theta$ of this antenna, then these two targets should be resolved, otherwise only one target will be detected.

The radar receiver adds together all the waves picked up from each element. Taking into account the different phases of these waves from each array element, we can analyze this process mathematically and plot the total power reflected off target 1 back to the antenna. This plot is shown in figure T5.2, assuming that target 1 is on boresight ($\theta = 0°$) and target 2 is offset from target 1 by 2° (so that $\beta = 2°$). If the array element spacing of figure T5.1 is $d = \frac{1}{2}\lambda$ and there are 57 elements in the array, then from note T4 you see that the antenna resolution is $^2/_{57}$ radians or 2°, so that this antenna should (just) be able to resolve our targets. This is confirmed in figure T5.2. The power from each target is plotted as a function of bearing angle θ. Note that the radar signal processor does not get to see these individual power plots; it is given only the sum of them by the receiver antenna. Receivers simply soak up the radiation that they were designed to detect, add it up, and pass on the resulting electronic signal to the signal processor. The job of the signal processor is to interpret this information, and it sees two separate peaks, in this case. So it knows that there are two targets—they are resolved. Now consider what happens if there are only 35 array elements in the receiver. This smaller antenna has an azimuth resolution of about $^2/_{35}$ radians, or a little over 3°. Now the summed signal has only a single (broad) peak—the signal processor interprets this as a single larger target. No resolution. This is one reason why the CH radars, with their wide mainlobes, underestimated the number of enemy planes.

T6. Amplitude Monopulse

Each CH radar receiver consisted of two mainlobe beams, offset in angle as indicated schematically in figure T6.1. A target is detected via both beams, but its apparent size (the echo power) differs. The beam that points more nearly at the target will hear a louder echo than the other beam. In figure T6.1 the ratio of powers P_1/P_2 is just the ratio of the lengths of the lines shown (as explained in note T3). The point is that this power ratio changes with target bearing. For a well-designed pair of beams, the ratio is a unique function of angle, so that a measurement of power ratio immediately provides an estimate of target direction angle, either azimuth or elevation.

The accuracy of estimation angle attained by this technique is limited by the accuracy of power ratio measurement. In practice, though, the results are very satisfactory. Instead of the target direction estimate being accurate to within a beamwidth, with amplitude monopulse this accuracy can be improved by a factor of 10 or 20 or 30, depending on the radar receiver. So this signal processing technique provides the best of both worlds: the beamwidth is wide enough so that the receiver can hear targets over a broad swath of sky, and yet the target direction can be estimated with high precision. One reason for the success of the amplitude monopulse approach is that it does not depend on the target power. Indeed, the target power will vary with range, and with other factors that I will discuss in chapter 2. In fact, it can fluctuate quite quickly. This makes no difference to the monopulse technique, however, because the ratio is determined by two measurements taken at the

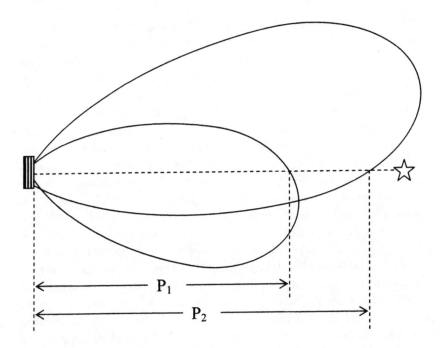

Figure T6.1 A *receiver* (left) *with two mainlobes detects a target (star). The echo power from the two beams is* $P_{1,2}$. *The ratio* P_1/P_2 *is a unique function of direction angle and is independent of target echo fluctuations. This amplitude monopulse technique is a robust method for improving angular resolution.*

same instant of time. The power levels in each beam may vary, but the ratio depends only on direction angle.

T7. Bandwidth

It takes only a short sentence to define the fundamental concept of bandwidth, but a couple of pages to show why it is important. The bandwidth of a receiver is simply the spread of frequencies that the receiver is able to detect. Thus, the human ear is sensitive to frequencies from about 20 Hz to about 20 kHz, so the bandwidth of our acoustic receivers is about 20 kHz (for a healthy young person). As we age, the upper limit decreases, and so our audio bandwidth is reduced. If my ears become damaged, so that they are sensitive only to sounds above 400 Hz and below 1,500 Hz, then my bandwidth is 1,100 Hz. The idea of bandwidth also applies to EM signals. If the spread of frequency components in a signal is large, then that signal is *wideband* or *broadband*. A tuning fork emitting a single pure tone is *narrowband*.

There are two reasons why the bandwidth concept is important: jamming and information content. In the restricted context of military signal processing (radar or sonar) a receiver must guard against *jamming*. I discuss jamming in chapter 4, and in this note will simply state the basic idea. A jammer is an enemy transmitter that sends loud noise into your receiver to mask any real signal that your enemy may be transmitting. Suppose you wish to eavesdrop on a conversation between two people some distance away. An ally of theirs wishes to deny you this information, and so broadcasts loud rap music at you through a megaphone, close by. (OK, so rap music is not the same as noise, but to a curmudgeonly engineer over fifty, such as myself, there is little difference.) He is jamming their signal by swamping your receivers with noise. Jamming is a significant part of electronic warfare (see chapter 4). The manner in which a receiver can defend itself against jamming is to tune in to only a narrow band of frequencies—those that are expected to contain the desired signal—and filter out everything else. This is because the power that a jammer is able to stuff into your receiver depends on the receiver bandwidth. The wider the bandwidth (i.e., the more frequencies to which the receiver is sensitive) the more jammer power intrudes. So, the CH radar receivers were provided with the capability of blocking out all but a 50-kHz band of frequencies. These narrowband receivers were less vulnerable to German jamming signals than were wideband receivers.

So why should receivers be wideband at all? This brings us to information content. Signals may be complex, in particular, if they originate from a complex source (such as the radar reflections from a large number of airplanes). A complex signal contains a lot of information—this makes the signal wideband.—Think of the amount of information necessary to *specify* a wideband signal (all those frequency components, each with their own amplitude and phase) and you can see that the information content of a signal increases with bandwidth.—So, when CH operators wanted to interpret a complex signal, they were obliged to let their guard down (widen the receiver bandwidth, so making it more vulnerable to jamming) so that all the frequency components of the signal could be detected. Skilled operators could then analyze the signal and learn about the airplanes that gave rise to it.

I can illustrate the problems that arise if we try to listen to a wideband signal using a narrowband receiver, as follows. In figure T7.1 I have a simulated signal, containing many frequency components (real signals are far, far more complicated than this, and they fluctuate in time, to boot, so this simple example is not realistic, but serves the purpose here). If this signal is detected with a receiver that has a bandwidth narrower than the signal bandwidth, then the received signal is distorted, as shown. Narrow the receiver bandwidth further, and the signal becomes more distorted. In particular, note how the fine detail has become smeared out and lost.

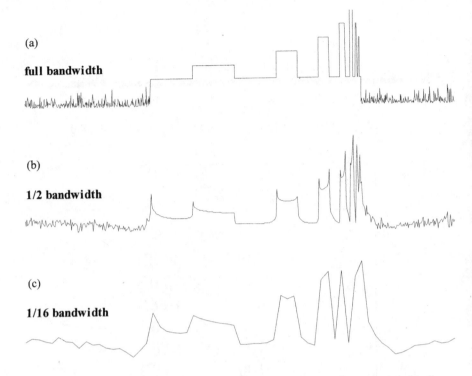

(a)

full bandwidth

(b)

1/2 bandwidth

(c)

1/16 bandwidth

Figure T7.1   (a) A synthetic signal amid noise. (b) If we try to represent this signal by utilizing only half the original bandwidth, the signal becomes distorted. (c) With 1/16 of the original bandwidth, the signal has lost much of the fine detail.

It is perhaps worth emphasizing that the jamming consequence of receiver bandwidth is pretty much restricted to military application, whereas the information consequence is universal—it applies to all signals.

Chapter 2. Remote Sensing Foundations

T8. Decibels

A *decibel* (dB) is a measure of the ratio of two numbers. The *deci* part refers to 10 (as in *decimal*), whereas *bel* honors Alexander Graham Bell, the founder of Bell Labs (where dB was developed in the 1920s for the telecommunications industry). In particular, the dB measures the ratio of two numbers on a logarithmic scale, as indicated in figure T8.1. We can describe large numbers by the number of digits needed to write them down. For example, in the figure you see a line with powers of 10 strung along it, from 1 to 1,000,000. Next to each number is the number of zeroes needed to describe each power, so 1,000,000 is labeled with the number 6. A dB is simply 10 times the label

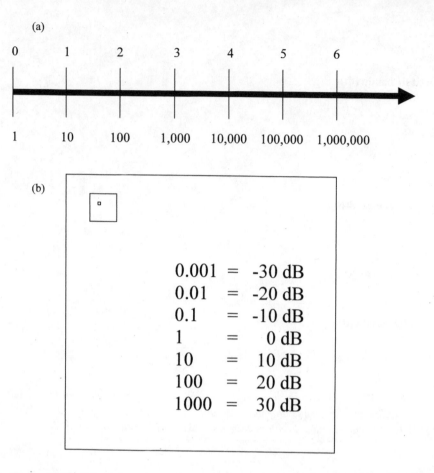

Figure T8.1 *Decibels: a logarithmic scale to handle very large and very small numbers. (a) Representing big numbers by the number of digits needed to specify them or, as here, by the number of zeros. (b) Squares of different sizes illustrate how easily big numbers arise: the biggest square has an area that is 10,000 times that of the smallest. Also shown, a quick conversion table from numbers to their decibel equivalents.*

number. Thus 1,000,000 is 60 dB. You can see that this is a very convenient way to label large numbers, and in the radar and sonar business we need large numbers.

Negative labels apply when the numbers are less than one, as shown in the figure. Those of you who are familiar with logarithms will know that all numbers, not just powers of 10, can be given dB labels. Thus 2 is approximately 3 dB, and 4π is about 11 dB — these particular values are engraved on the hearts of all radar system engineers. If logarithms are not your thing, then simply consider decibels to be a convenient gauge of scale. So, 17 dB is between 10 dB and 20 dB, and so represents a number bigger than 10 but less than 100 (in fact, 17 dB equals 50, pretty nearly). Similarly -23 dB represents a number smaller than $^1/_{100}$ but bigger than $^1/_{1000}$ (in fact, it is $^1/_{200}$).

To appreciate the usefulness of dB, consider the three squares in figure T8.1. The biggest square has an area that is 100 times the area of the medium square, which in turn has an area that is 100 times the smallest square. Thus, the largest square is 40 dB bigger than the smallest. Consider now a cube—say a salt crystal—that is one-tenth of an inch (0.1 inch) along each side. You will have no difficulty imagining a string of 1,000 such salt crystals placed next to each other along a straight line—picture a thin white line on your living room floor, 8 feet 4 inches long. Now imagine that 100 such lines lay side by side, forming a white square on your floor. There are 1,000,000 salt crystals in this square. Put another way, the size of the square is 60 dB relative to one crystal. Now imagine placing 100 such squares one atop the other, to form a cube that is 8 feet 4 inches on a side (you have a big living room). The size of this cube is 90 dB relative to a single crystal. It represents a scale factor of one billion (1,000,000,000), which is about as much as my brain can envisage. But nature requires us to grasp much larger scales. Suppose that an extraterrestrial civilization of little green men has, for reasons best known to itself, constructed a gigantic cube of salt crystals out in space. A cube of this size would just be big enough to contain the earth. So how many crystals does it contain? Well, about 400,000,000,000,000,000,000,000,000,000. This is a big number, perhaps too big to write down conveniently. On the other hand it can be expressed as 296 dB, which is easy to write down, and more readily conveys the scale of the cube.

Note that dBs need a reference point, since they refer to the size of something relative to something else. For our cubes, the dB scale was referred to a salt crystal. In acoustics the dB refers to the lowest level of sound that the human ear can perceive. So when you are told that normal conversation is about 60 dB, you know that the intensity of sound in a normal conversation is about one million times the intensity of the smallest sound you can detect. A loud noise (such as thunder) is typically 85 dB. Permanent ear damage occurs with sounds around 120 dB. So, the range of sound intensities that we can hear varies by a factor of one trillion (this is called the *dynamic range* of the ear). Radar systems engineers often refer to electrical power in units of mW (one-thousandth of a watt), so that a 30 kW (30,000 watts) transmitter consumes about 75 dBm power. The "m" refers to milliwatt. If the echo power detected by a radar receiver is less than the transmitted power by a factor of 5,000,000,000,000, then we say that the received signal is -127 dBc, meaning 127 dB below the transmitted (carrier) signal power.

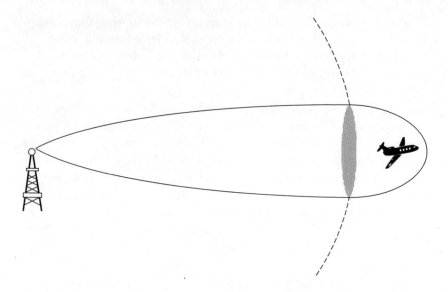

Figure T9.1 *If a transmitter sends microwaves in all directions, the power density at distance R (dashed line) will be the same, whatever the direction. If the transmitter concentrates its power in one direction (the mainlobe beam is shown) then most of the power at distance R is concentrated in the shaded area. Antenna gain is the ratio of the surface area of the sphere to that of the shaded area.*

Armed with this essential engineering lore, you can now impress people at parties, by pretending to be the sound engineer for a famous rock band. I wish you more success at that game than I ever enjoyed.

T9. Antenna Gain

We saw in notes T3 and T4 that radar beams are directional, and that the radar antenna engineer can, to a large extent, control the degree of directionality. He (and in my experience all antenna engineers, though not all radar engineers, are male) chooses the antenna dimensions and these, relative to transmitted wavelength, determine beamwidth. In figure T9.1 I show a simplified radar beam that is pointing at an airplane target. The airplane is *on boresight,* meaning right in the middle of the antenna mainlobe. Because of the antenna directionality, most of the transmitted radiation is pointed at the airplane, illuminating it with microwaves, which then cause a strong echo to return to the radar. Now suppose that the directional antenna is replaced by a second, reference antenna that transmits the same power in the airplane direction but does not have a directional beam. The microwave radiation is transmitted in all directions

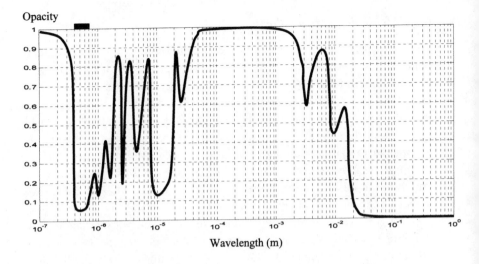

Figure T10.1 *The fraction of EM radiation that is absorbed as it travels through one mile of atmosphere (opacity) is here plotted as a function of wavelength. The wavelengths of visible light are indicated by the bar, at top. Much of the infrared (IR) spectrum (from about 10^{-6} to 10^{-4} meters) is strongly absorbed, though there are windows. IR "heat-seeking" guidance systems make use of these windows. In the microwave region (the right half of the graph) absorption is variable: strong at millimeter wavelengths, weaker at centimeter wavelengths, and practically nonexistent at metric wavelengths. This is why long-range radars use longer wavelengths.*

uniformly (the mainlobe reduces to a sphere centered on the antenna). This reference antenna must send out a much higher level of microwave power to hit the plane with the same power as our directional antenna. Our directional antenna is said to have a higher *gain*, and this gain is measured relative to the reference antenna. In figure T9.1 I show part of a sphere (the reference antenna beamshape), intersecting the directional antenna mainlobe. The intersection area shown contains half of the transmitted power—the rest is spread over the outer part of the mainlobe and the sidelobes. The directional antenna gain is simply the ratio of the sphere area to the shaded intersection area. Because it is a ratio, gain is conveniently expressed in dB. For a typical airborne microwave surveillance radar antenna, gains are in the region of 30 dB or so. Tracking radars and ground-based (or ship-based) radars can have bigger antennas and hence larger gains. Antenna gain is simply a geometrical factor, reflecting the fact that transmitter power is confined to within a small part of a spherical shell—another way of saying that the antenna is directional.

T10. Atmospheric Attenuation

In figure T10.1 we can see how the earth's atmosphere absorbs EM radiation. Atmospheric opacity is plotted against wavelength, for wavelengths varying from ultraviolet (10^{-7} m) to metric (1 m). Opacity is simply a measure of the fraction of radiation that is cut out per kilometer or per mile traveled through the air. For visible light the opacity is small—light can travel many miles through the air before it fades away to nothing. This is hardly surprising: animals and humans would not have evolved vision over a bandwidth of the EM spectrum that was rapidly attenuated. By way of contrast, much of the infrared spectrum is stopped dead by the air. The microwave region shows virtually no attenuation of EM intensity above 10-cm (0.1-m) wavelengths, which explains why long-range radars are of metric and centimetric wavelengths. Millimetric waves, used by tracking radars because of the high angular accuracy they provide (even with relatively small antennas) are sopped up quite severely, however. So mm-wave radars have only a short range. This is not so important if the radar is designed to track incoming missiles—for example, sea-skimming missiles approaching a warship. In that case high resolution is more important than long range. You can see from the figure that certain "windows" exist in the mm region—corresponding to frequencies of about 35 and 90 GHz. So, if you see a small tracking radar antenna atop a warship mast, then you can be pretty sure that the operating wavelength is in one of these two narrow bands. Even so, the atmospheric attenuation limits range to perhaps 10 or 20 km. To summarize: the vast variation in radar ranges (from a few kilometers to many thousands of kilometers) is largely due to the attenuation of EM radiation by the atmosphere. (Specialized ground-penetrating radars, used by archaeologists and treasure hunters, have ranges of only a few meters. Of course, this is because the ground cuts out EM radiation much more severely than does the atmosphere, at all wavelengths.)

Atmospheric attenuation happens for one of two reasons. Either the radiation is absorbed by atmospheric gases (principally oxygen and water vapor) as in figure T10.1, or it is scattered by rain or by aerosols or dust. In figure T10.2 you can see the effect on microwaves of oxygen and water vapor absorption, over a much narrower portion of the spectrum than in figure T10.1. The attenuation is expressed as decibels per mile in the United States, whereas European radar engineers use decibels per kilometer.

There are some odd mixtures of measurement units in the radar business. For example, range is often expressed in kilometers, whereas altitude is given in feet. Even worse, sometimes both English and metric measures are mixed up in the same unit, as when altitude is expressed in "kilofeet." One day we will all agree on a standard. Meanwhile, I am as guilty as everybody else—

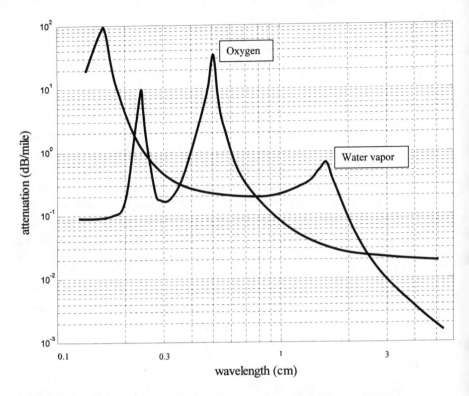

Figure T10.2 *For a narrow portion of the electromagnetic spectrum in the microwave region, the two-way attenuation of power due to atmospheric oxygen and water vapor is shown.*

note in this book that I usually adopt metric units but, in deference to U.S. readers, I sometimes state lengths in miles, feet, or inches.

The amount of oxygen in the atmosphere is fairly constant, and so the attenuation figure for oxygen does not vary from place to place, except with altitude. But water vapor is a different kettle of fish, so to speak. Water vapor concentrations peak at low altitude over the sea, in clouds, and above rain forests, but they tend to be low over deserts and in the upper atmosphere. This variability is a source of considerable uncertainty for the radar engineer who wants to account for atmospheric attenuation in his radar range equation calculations. The problem is exacerbated when scattering particles are present. Pollutants, smoke particles, and other suspended solids cause EM radiation to scatter. The scattering is wavelength dependent, but in a way that is completely different from that observed for absorption. It varies considerably over the globe, because aerosol and airborne particle concentrations vary considerably. The phenomenon of EM atmospheric scattering and its variability is familiar to us from everyday observations of sunlight. The sky is blue and sunsets are red because of scattering. In a polluted environment or following a large volcanic eruption (which throws vast quantities of small dust and smoke particles up into the air, and around the globe) the sunsets can become much more vivid.

It is hopeless to try to account for all this variability, so the radar (and sonar) engineer lumps all sources of attenuation together to come up with an average *loss factor*, here denoted L. This loss factor is usually expressed in decibels per kilometer. If the range from radar to target is known, then it may be expressed simply in decibels. So, if $L = -7$ dB then 80 percent of the radar transmitted power gets lost on the way to or from the target, due to atmospheric absorption or scattering. In particular scenarios, detailed loss factor calculations may be needed, and the radar textbooks are full of graphs showing us how to incorporate such effects as rainfall (loss depends on rainfall rate, as well as the distance through rain that microwaves must travel), snow and hail, variation of air density with altitude, and so forth. I will spare you the details and simply include a factor L in our range equation. You know now that L depends on atmospheric conditions, that average losses can be determined assuming certain meteorological conditions, but that the loss factor estimates may vary because of localized effects.

T11. Radar Cross Section

A simple target, such as a metallic sphere, looks the same no matter from what direction you stare at it. Because of this, the amount of microwave power that it reflects does not depend on look angle—the sphere is said to be an *isotropic reflector*. Now it is a straightforward matter to define the radar cross section of such a reflector. The RCS is simply the area of the sphere that is seen by the microwave radiation.

Purists might curl up their toes at my definition of RCS. In fact, the area of a sphere that is seen by microwaves is not necessarily the geometrical area of the sphere. If the size of the sphere is similar to the wavelength of the microwave radiation, then RCS can differ significantly from geometrical area. This is a characteristic of EM radiation in general, and not just microwaves, but it can be ignored here. For the most part, radar targets of interest are much bigger than the wavelengths of the transmitter radiation that illuminates them (for microwaves, recall, the wavelengths are of the order of centimeters). In this limit, the radar cross section of a sphere and its geometrical area are pretty much the same thing. So a spherical target has the same RCS no matter what angle it presents to a radar receiver. A sphere with a cross section of 2 m^2 has an RCS of 3 dBm^2. Recall that dB is a ratio, requiring a standard reference, in this case a reference area. For RCS the reference area is taken to be that of a sphere with cross-sectional area of 1 m^2.

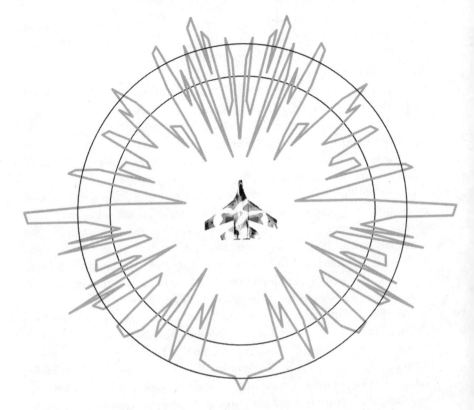

Figure T11.1 *Schematic illustration of radar cross-section variation with aspect angle. The distance from the airplane center represents radar cross section in decibels. Scale is indicated by the circles, which are separated by 5 dB.*

For nonspherical targets, RCS is more complicated. Consider the fighter plane of figure T11.1. The area that it presents to you depends on the angle you look at it, and the same is true for the plane's RCS. In figure T11.1, I show a schematic representation of how a fighter plane RCS varies with look direction. The two circles represent RCS in decibels, with the larger circle corresponding to an isotropic reflector that is 5 dB bigger than the smaller circle. So you see how much the RCS of a complicated target such as a fighter plane can vary with angle.

This figure is a schematic—it does not correspond to the RCS figures for a real plane, but the spiky shape is typical. I cannot show you the equivalent figures for real fighters because these numbers are classified. They are secret because, of course, the radar RCS of a target tells an enemy how easy it is to detect the plane via their radars. A lot of effort has been put into the measurement of military airplane RCS, to minimize the cross section, so that the plane becomes more difficult to detect and track. To measure RCS, the plane is placed on a giant pedestal and rotated. While this happens, the plane is being illuminated by a narrowband transmitter, with a well-defined frequency, and the echo power received at a specified angle is noted. To eliminate reflections from other sources, the area behind the plane is covered with RAM (radar-absorbing material). You can appreciate how cumbersome and complicated such RCS measurements are when you consider that the plane must be rotated in three dimensions to get a complete picture—the RCS of figure T11.1 assumes that it is viewed sideways—and that RCS depends on radar frequency and look direction.

Given that the cross section of a complex target strongly depends on look angle, how should we define the RCS? It is standard practice to consider the average value, over all angles, and specify this value as the target RCS. This averaging process is equivalent to replacing the complex target by a sphere, and saying that the target RCS is the (constant) cross section of the sphere.

The fact that manufacturers go to such lengths to measure the RCS of military equipment shows the importance of RCS. During the procurement process for a military radar system, the buyer will request several bids from different radar designers, and will specify the radar performance. This performance will typically be expressed as follows: "Your radar must be able to detect a target with 1-m^2 RCS at a range of 20 miles with 90 percent probability." I will get into the probability bit soon enough, but for now you may note that RCS permits radar performance to be specified precisely, so that the buyer knows what he is getting when he orders a radar system. This is not the least important reason why RCS matters.

The main military reason why RCS matters, as mentioned, is to minimize target visibility to enemy radars. To this end military airplanes and ships are designed with RCS in mind. A plane will typically have an external shape that is determined by aerodynamics but, for stealth bombers, the aerodynamics are traded for low RCS. Note the weird appearance of a B-2 Spirit stealth bomber (fig. T11.2). The angular edges are designed to reflect radar transmissions away from the ground-based radar receiver. All features that may contribute to increasing cross section, such as air intakes and jet exhausts, have been relocated to the upper side of the airframe, so they are not seen by radars on the ground. (Hiding the exhausts in this way also renders the B-2 much less visible to infrared heat-seeking missiles and ground-based infrared-imaging equipment.) The engines have been brought inside the airframe, so they cannot be detected by radar. The control surfaces (wing flaps) are flush with the wings and, again, appear to be on the upper surface of the wings rather than hanging off the back, as with conventional civilian airliners. In addition to these structural changes, the B-2 skin is made from specially adapted low-RCS composites and painted with RAM material that ab-

Figure T11.2 A B-2 *Spirit stealth bomber of the 509th Bomb Wing over the Pacific Ocean. Stealth design trades aerodynamic efficiency for very low RCS. Note that engines are internal, with exhausts on the upper wings. The straight-edged wings and special coating help to reduce radar visibility. U.S. Air Force photo by Master Sgt. Val Gempis.*

sorbs EM radiation. All these modifications must make the B-2 fly like a brick, which shows how much importance is attached to low-RCS design. The cost of one of these bombers is staggering: $1.157 billion, or $215 per ounce (77 cents per gram, in metric units). I obtained this information from a U.S. Air Force Web site, which provided much helpful information about this impressive bomber, but nowhere on this Web site will you find the RCS.

It is rumored that stealth bombers such as the B-2 have an RCS of well below a square inch, perhaps as low as a square millimeter. I attended a radar course in London, in 1988, given by David Barton, an American radar guru well known throughout the international radar community. At this time it was rumored that the United States was developing, in great secrecy, a stealth aircraft. Given the importance of RCS to radar designers, this rumor caused considerable interest, and one German delegate asked Barton, with a bluntness that may surprise some readers: "What is the RCS of the secret stealth bomber?" Barton was perhaps familiar with fielding such questions, judging by his answer: "Well, you know, the Air Force denies the existence of such a project and so, even if such a stealth bomber existed (which it doesn't) and

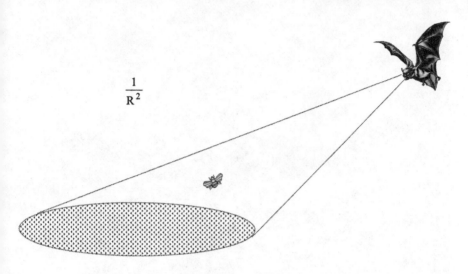

$$\frac{1}{R^2}$$

Figure T12.1   *Type 2 (surface) clutter: the bat receives echoes from the ground. These vary inversely as the square of the range.*

I knew the RCS (which I don't) I wouldn't be allowed to tell you." For comparison, note that a conventional bomber of the same size would have an RCS exceeding 20 m² — say 31,000 square inches. So, the huge development effort and cost have resulted in a plane that is easy to see but is all but impossible to detect on radar, and so is virtually invulnerable to attack if flying above the clouds or at night. (As a radar designer I worry what happens when a stealth bomber opens its bomb-bay doors — the RCS may spike — but I guess they must have thought of that.)

T12. Clutter

In figure T12.1 we see an example of type 2 clutter. Here an MB800 airborne interceptor (I'm going to have to stop referring to bats in this way) is transmitting acoustic radiation onto the ground. It does not mean to do this — it really wants to illuminate (I will avoid using the awkward equivalent term for acoustic beams: *ensonify*) the air in front of it to detect flying insects. However, much of the bat's acoustic power travels beyond the prey and onto the ground below, where it is reflected back into the bat's ears. A similar problem exists for fighter planes diving down on bomber targets, or for ground-attack aircraft. Also submarines that transmit a sonar beam upward may suffer from the confusing effects of reverberation echoing off the sea surface. All these examples illustrate surface clutter, here type 2. You can see why I have so labeled it: the range behavior goes like R^{-4} (range equation) multiplied by R^2 (surface area illuminated) to yield R^{-2}. So surface clutter RCS (and

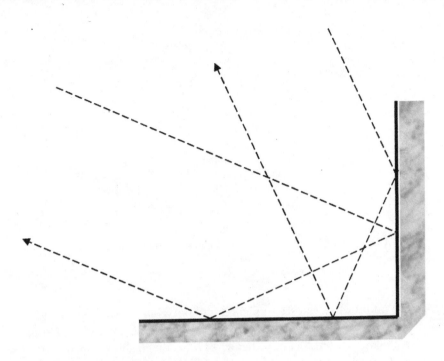

Figure T12.2 *Corner reflectors send radar signals back toward the radar, and so have a large RCS over a wide range of angles.*

hence surface clutter power) falls away as the inverse square of range. The units used for measuring surface clutter density are, bizarrely, square meters per square meter, m^2/m^2.

Consider the enormous variety of surface clutter. (How many different surfaces can you imagine?) Grass and pavement will be common surfaces for bat echolocation clutter. Prairies, hills, deserts, urban buildings, forests, ice fields, and oceans all reflect clutter back into an airborne radar receiver. The ocean bed and surface reflect sonar power into a submarine sonar receiver. Each of these surface types presents its own unique, and sometimes very troublesome, properties. Two examples. Urban areas often present right angles (for example, between building walls and pavement), which leads to the phenomenon of *corner reflection*. As you can see in figure T12.2, a corner reflector will send back a lot of clutter power preferentially in the direction from whence it came, that is, back toward the radar. So radar images of urban areas often contain bright flashes. (Another problem that arises from man-made clutter is that certain common conductors generate large clutter power. For example, when illuminated by just the right [i.e., wrong] wavelengths, utility lines and chain-link fences seem, to a listening radar receiver, to sing out loud.) These flashes do not necessarily originate from flat surfaces seen

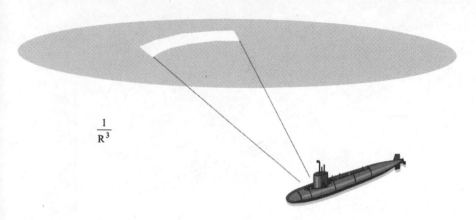

$$\frac{1}{R^3}$$

Figure T12.3 *Type 3 clutter: surface clutter restricted to a narrow range. Here acoustic radiation from submarine sonar is reflected off the underside of the sea surface.*

square on (as when you see sunlight flash from a distant window)—they are much more commonly due to corner reflectors. A situation similar to the two-dimensional illustration of figure T12.2 exists in three dimensions, as when two walls and a pavement intersect at right angles. My second example is sea clutter. The ocean surface changes shape and moves, and the clutter returned from it to a radar receiver is particularly complex. It varies in time on two timescales: froth, foam, and spray fluctuate each millisecond, whereas the large ocean waves change over a timescale of seconds. Your author is one of several hundred radar systems engineers around the world who has spent a significant fraction of his career investigating just this one problem. It was militarily important during the Cold War, when spotting submarine periscopes from a distant aircraft was deemed to be a tactical imperative.

Type 3 clutter is illustrated in figure T12.3. I have chosen a submarine sonar here, but might equally well have picked an airborne radar or bat or whale echolocator. The sonar in figure T12.3 is interested in a particular range—determined, as we have seen, by time delay of the transmitted signal—and so filters out everything else. The sector shown has an area that is proportional to R, so that if the submarine wants to investigate a sector that is more distant, then that sector will be proportionally bigger. Thus here the clutter RCS behaves like R^{-4} (range equation) multiplied by R (clutter area), that is, like R^{-3}. So type 3 clutter is really a subset of surface clutter, because it depends on reflections from a surface.

Some clutter falls between the types that I have described. For example, trees may be considered as both volume clutter and surface clutter, so that

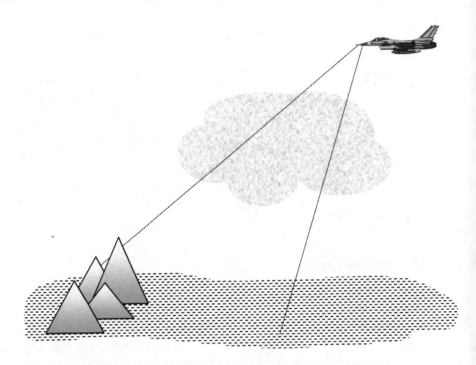

Figure T12.4 *Real clutter environments can be complicated, in particular, for air-borne radar, with several different clutter types contributing to the radar echo.*

the way tree clutter depends on range will be complicated. Also, many environments contain clutter of different types. This point is readily grasped by glancing at figure T12.4. An airborne radar beam penetrates a cloud (type 1) to illuminate hills and plains (two varieties of type 2). So, the manner in which clutter power varies with radar range is complicated, in general, and depends on details of the environment. All of this clutter must be recognized as such by the radar/sonar signal processor, and must not be mistaken for a target. In chapter 3 I show you how signal processing discriminates between clutter and targets, suppressing clutter and magnifying target echoes.

One last complication of clutter is illustrated in figure T12.5. Acoustic radiation from a submarine sonar transmitter is reflected from the sea surface and from the seabed. The complicated multiple reflections depend on initial direction of the radiation—two rays are illustrated—and the echo signal in this case will be very complicated. I won't say any more about this phenomenon here, because it more properly falls under the heading of anomalous propagation, which is best deferred to a later chapter. Suffice it to say that the scrambled clutter echo can be utilized by an intelligent target to avoid detection.

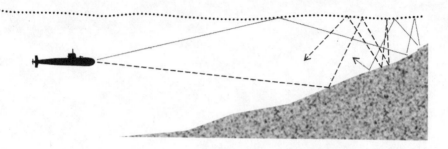

Figure T12.5 *Sort this out. Sonar transmissions are reflecting off both surfaces of a shelving beach. This will look messy and complicated by the time it gets back to the sonar receiver. A potential target may hide in this mess.*

T13. Statistics Primer

Four main concepts (plus a couple of auxiliaries) from the large and expanding world of statistics will suffice for our purposes: *probability, mean value, standard deviation*, and *sampling*. These words are fairly familiar ones, but their meanings are often appreciated only vaguely. Understand these concepts—and here I hope to convey them in a manner more palatable than cod liver oil—and you can appreciate many of the key aspects of remote sensing without delving into mathematical detail.

Probability is simply frequency of occurrence, based on past experience. Thus, if you flip a coin and it comes up heads 507 times out of 1,000, then you say that the probability of obtaining heads is $^{507}/_{1000}$ or 0.507. If you roll a pair of dice 6,000 times and the two numbers add up to seven on 987 occasions, then the probability of throwing seven is $^{987}/_{6000}$ or 0.1645. Note that probability always lies between zero and one.

Now you return to the coin, and patiently flip it another 1,000 times. On this occasion you obtain heads 495 times so that, based on your last 1,000 flips, you estimate the probability of heads as 0.495. Which is right, 0.507 or 0.495? They are both right, based on their set of 1,000 samples. But you could add together the two results to obtain a probability of $^{1002}/_{2000}$ or 0.501. This value is a better estimate of the "true" probability of turning up heads for that particular coin, because it is based on more experience. If you were to flip the coin 25,000,000 times (don't try this at home, folks, life is too short—I will soon get my computer to do it for you) you might obtain a probability of 0.5006 or 0.4997 or something very similar. If your coin is a fair one, then theory and common sense agree that the "true" probability is exactly one half, but to prove it you would have to flip the coin an infinite number of times. Indeed, my definition of probability *requires* that you flip the coin an infinite number of times, because only then can you acquire *all* the possible experi-

Number of trials

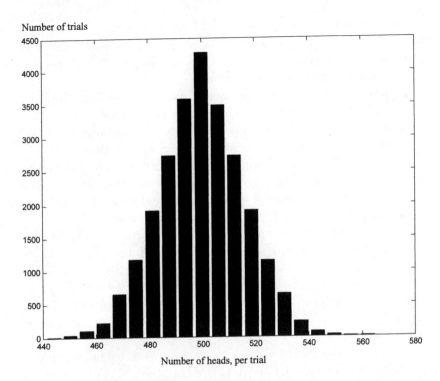

Number of heads, per trial

Figure T13.1 *The coin-flip distribution. In this case the coin was flipped 25,000,000 times, and for each trial of 1,000 flips the number of heads is noted. About half the trials lie within one standard deviation of the mean value. In this case, because the distribution is symmetric, mean value is in the middle, at the peak. Divide this distribution by the number of trials, and we obtain the probability density function for coin flips.*

ence that can be obtained for this coin, but this is impossible to do. Clearly, the more you flip the coin the closer you will get to the "true" probability, but any realistic measurement of probability is only an approximation.

A mean value is simply an average value (technically you may hear people insist on calling it the "arithmetic mean" but that is just a fancy name for what we all understand by "average"). So when I combined the estimates of probability for my first two sets (*trials*) of 1,000 flips, I obtained the mean value of these two results. Now I will ask my computer to perform 25,000 trials of 1,000 flips, and plot the 25,000 estimates of probability for obtaining heads. Voilà, figure T13.1. If my computer averages all 25,000 estimates, it comes up with a better estimate of the probability for flipping heads, in this case about 0.49996.

The concept of standard deviation is also on display in figure T13.1. The graph of estimates shows values cluttered about the mean value. More estimates are close to the mean value than are further out in the wings. We have a *distribution* of estimates. We can characterize how meaningful our mean estimate is by saying how much spread there is among our 25,000 estimates. After all, if the estimates of probability are spread out over a large value—say between 0.001 and 0.997—then any single estimate of probability (based on one trial of 1,000 flips) is pretty meaningless. On the other hand, if our 25,000 estimates are spread out over a very narrow range, say between 0.498 and 0.501, then we can be pretty sure that our mean value is close to the true value, because the individual estimates agree among themselves. The amount of spread is called the standard deviation, and for the distribution of figure T13.1 it is 0.01585. We express the mean value and standard deviation together as 0.49996 ± 0.01585 (read "0.49996 plus-or-minus 0.01585").

There is an old joke about the statistician who drowned in a river of average depth three feet. If there was such a person, then he was a bad statistician. He took heed only of the mean value of depth estimates, and did not take into consideration the standard deviation. Had the river depth and standard deviation been 3 ± 1 feet, he would likely have lived but, unfortunately for him, this river was 3 ± 5 feet deep, and so he drowned—a victim of misunderstood statistics.

The shape of the graph in figure T13.1 can be calculated theoretically by statisticians, assuming an infinite number of trials instead of our 25,000 trials. We need not be concerned with the details, but note that the shape for flipping coins is not the same as the shape for throwing dice. In general, the shape of a distribution depends on the statistical process. If we add up all the bars in the graph, the total comes to 25,000, of course, because that is the number of trials we conducted. If we now divide the distribution of figure T13.1 by the number of trials, then we have the *probability density function* or PDF for flipping coins. In this book I will rather loosely use the terms PDF and distribution interchangeably. Both terms describe the likely outcome of a statistical measurement. Some statisticians would object vehemently to my blurring the distinction between PDF and distribution, but we cannot hear them for the *glug-glug* sound as they slip below the surface of the river.

Now, I will leave the realm of general statistics and focus more specifically on signal processing to give you an idea of statistical sampling. In figure T13.2 we have a wiggly line representing a signal returned from the environment into our radar or sonar receiver. This signal may be a target echo or it may be a clutter echo—we don't know which, because this signal has not yet been processed. When the train of transmitted pulses was sent out it had a known waveform, but it has now returned to us in a mangled and barely recognizable form. To begin the process of classifying this particular echo signal, we sample it. The signal processor deals in voltages, and the wiggly line of fluctuating voltage is our signal. The signal is sampled at short intervals

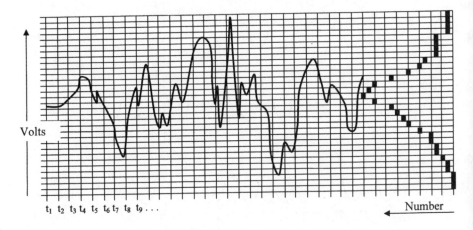

Volts

t₁ t₂ t₃ t₄ t₅ t₆ t₇ t₈ t₉ ...

Number

Figure T13.2 *Sampling of a waveform at regular time intervals. For a remote sensing system, the waveform is the stream of echo pulses returned from the environment. On the right you can see the distribution of voltages for this waveform—this is proportional to the number of times the waveform passes through a given voltage level. Turned on its side, we have a distribution similar to that of figure T13.1.*

$t_1, t_2, t_3,$ . . . as it flies along from the reciever to the signal processor. The sample rate will be much greater than the pulse repetition rate, so that each pulse is sampled many times. The many samples that make up our signal form a distribution of voltages. This distribution is shown in figure T13.2; it shows how likely it is for a randomly chosen sample to be at a particular voltage. From the last few paragraphs you can see that there is a mean value of voltage for our signal, and a standard deviation. Note that the distribution in this particlar case is asymmetric, unlike the distribution for coin flips, in that the left wing is longer than the right wing.

The distribution of samples depends on the signal statistics. Now that you are armed with the main concepts, we can tackle the statistics of radar and sonar signals.

Chapter 3. Signal Processing Techniques

T14. Coherent Radar

Phase coherence, to give its full name, appears ordinary enough but is vital for radar and sonar image formation and for Doppler processing—key processing techniques in the remote sensing locker. *Coherence* or *noncoherence* is a characteristic of the radar transmitter. (As distinct from *incoherence*, which is a property of some radar textbooks and manuals, and of all legal documents.) One limitation of the magnetron is that it is inherently noncoherent.

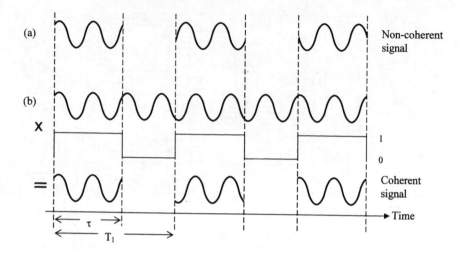

Figure T14.1 *Here are two transmitter waveforms that consist of a single frequency. (a) The noncoherent signal has no phase relationship from one pulse to the next. (b) The coherent signal is constructed from a single continuous wave that is modulated (multiplied) by a square waveform of ones and zeros, resulting in pulses (of duration τ and repetition interval T_1) with continuous phase.*

This means that adjacent, consecutive pulses of microwave energy that are emitted by the magnetron have a random phase. The basic idea is illustrated in figure T14.1. Pulses of duration τ possess a definite wavelength (about 9 cm in the case of the Randall and Boot cavity magnetron), but the wave phase at the beginning of the pulse bears no relationship to the phase at the end of the preceding pulse. Contrast this with a coherent power source, such as the later *traveling wave tube* (TWT) and *Klystron*. Also shown in figure T14.1, the coherent waveform maintains phase continuity from pulse to pulse. In fact the coherent waveform is much like a continuous wave, only modulated so that it is chopped up. Recall (note T2) that CW transmitters do not lend themselves to range estimation, and so to obtain range information we must pulse the transmitted power and estimate range by timing the pulses. On the other hand, continuous waves are needed for estimating target speed, as we will soon see, and long pulses estimate speed more accurately than short pulses do. So here we have a fundamental trade-off: we want short pulses for range estimation accuracy and long pulses for speed estimation accuracy. We can get *both* by emitting coherent pulse waveforms. Consider a *pulse train* or *pulse burst* of, say, $n = 100$ pulses (a burst of three pulses is show in figure T14.1). Each pulse has a duration τ and the whole train is of much longer duration $T_n = nT_1$. T_1 is the interval between pulses, usually called the *pulse repetition interval* (PRI). High range-estimation accuracy is obtained if τ is small, and (if the waveform is coherent) high speed estimation accuracy is obtained if T_n is long—the best of both worlds.

T15. The Doppler Effect and Doppler Processing

J. C. Doppler was an Austrian physicist who lived during the first half of the nineteenth century. He discovered that the pitch of a sound depended on how fast the source of sound was moving. The equivalent statement is true for light: the color of light depends on the speed of the source. In fact, it is a phenomenon of any type of wave with a source that moves relative to the observer. We are all familiar with the "ambulance effect" whereby the wail of an ambulance siren seems to lower in pitch as it passes you. The pitch is higher when the ambulance approaches than when it recedes. This is the Doppler effect.

A few comments may help to clarify some common confusions. The shift in the color of stars, due to their movement relative to earth, is indeed a Doppler shift. The same is true for distant galaxies, though the reason for most of the Doppler shift in this case is because the universe is expanding. The detailed understanding of cosmological Doppler shifts, and of all EM frequency changes that result from high-speed objects (comparable with the speed of light), requires an application of Einstein's Special Theory of Relativity. Understanding the Doppler shift for sound waves on earth requires the simpler (and more intuitive) Galilean Relativity utilized here. These two theories are not consistent: at high speeds they make different predictions, and Einstein wins. At low speeds, as here, they reduce to the same theory.

We can see how this effect arises by considering figure T15.1. Movement of a transmitter toward a stationary receiver (actually all that matters is the relative movement of transmitter and receiver) means that the frequency received, f, is greater than that transmitted, f_0:

$$f = \left(\frac{c + v}{c}\right) f_0$$

For a remote sensor, the signal has to travel both ways: from transmitter to target and back to the receiver. In this case it is the speed of the antennas (both transmit and receive) relative to the target that counts. The two-way Doppler frequency shift is

$$f = \left(\frac{c + v}{c - v}\right) f_0 \tag{3}$$

where c is the speed of the signal, either the speed of light (for radars) or of sound in seawater (for sonars) or of sound in air (for echolocators). The speed

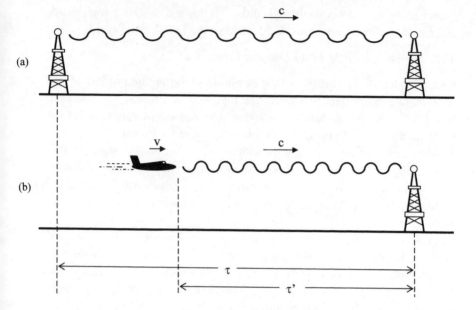

Figure T15.1 (*a*) *A stationary transmitter at left sends an EM or sound wave, consisting of a single frequency, to a stationary receiver. The length of the pulse* $L = c\tau$ *is chosen to be exactly the separation of the two antennas.* (*b*) *Now the transmitter is moving. It transmits the same length of signal but by the time it finishes transmitting it has moved. The receiver perceives the same number of waves as for the stationary transmitter, but now they are compressed into a shorter time:* $L = c\tau = (c + v)\tau'$, *so the frequency has increased.*

at which transmitter and receiver approach each other is v. v is usually much smaller than c (always, for radar sensing, since the speed of light is so big) and so we can approximate equation (3) as follows:

$$f \approx \left(1 + 2\frac{v}{c}\right)f_0$$

Now you see why the Doppler effect leads to a frequency shift. Since the speed of a wave is related to frequency and wavelength by $c = f_0\lambda$ we can write the receiver frequency as

$$f \approx f_0 + 2\frac{v}{\lambda} \tag{4}$$

so that the Doppler frequency shift is about twice the relative transmitter/receiver speed divided by transmitted wavelength. (If the antennas and the target are moving away from each other, then the frequency shift is negative; that is, the received signal has a frequency that is less than f_0.)

I hope it is clear from equation (4) how our remote sensor can estimate target speed: it measures the difference between the frequency it transmits to the target and the frequency it receives from the target, and then from (4) calculates target speed v. How accurately can a radar or sonar signal processor estimate v? That depends on how accurately it can estimate the frequency shift.

If we receive a single-frequency signal of duration T, then we can estimate that frequency with accuracy $\Delta f \approx 1/T$. So, the longer a signal lasts (the more time we have to gauge its frequency), the better is our estimate. If the signal is not a continuous pulse, but is a *coherent* train of n pulses, with the same total duration, this same accuracy obtains. This is important because good range accuracy requires short duration pulses. Now (cf. fig. T14.1) we can have short pulses *and* long pulse trains to give us both high range estimation accuracy and high frequency estimation accuracy.

The target velocity is estimated to an accuracy of $\Delta v \approx \lambda/2T$, where T is the pulse train length. (Police speed guns work on the Doppler principle. Originally these guns transmitted radar wavelengths, but over the past few years they have switched to laser wavelengths—much shorter and so, as you can see from the equation, much more accurate. Another benefit of the shorter wavelengths is that, for the same size gun antenna, the beam is much narrower, as we saw in note T4.) So, to take a typical airborne radar example, a 3-cm microwave pulse train of duration 1 ms is capable of estimating target speed to within about 15 m s^{-1}. Not bad. This estimation accuracy comprises the basic entry-level capability of a pulse-Doppler radar—in chapter 6 we see how to (greatly) improve on this accuracy.

T16. Video Integration

I can plot the results of my video integration simulation (fig. 3.6) in a different way. In figure T16.1 you can see the power distribution of the 100,000 samples. This method of presenting the data emphasizes the statistics of the noise, and of the target. (I have assumed that the target fluctuates "with Sw-1 statistics"—the most common type of radar target.) When the noise is integrated, the distribution changes. Here the mean noise power has been chosen to be one, in arbitrary units. When we integrate the "noise plus 3-dB target" echoes (i.e., the target has twice the mean noise power), we see that the distribution changes again. The mean power is now three, as we would expect (1 from noise plus 2 from target). The integrated "noise-plus 0-dB target" echoes have a mean power of two, as they should. Having performed the integration, we now need to apply a threshold, so that our signal processor can tell us whether it thinks a target is present in the echo pulses. Recall the

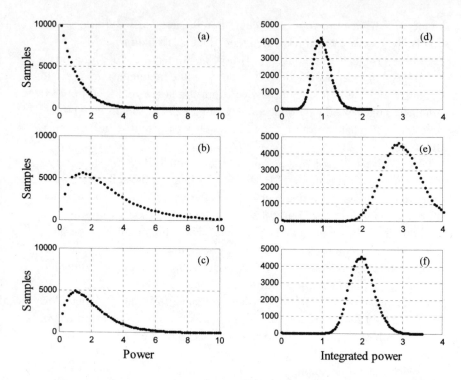

Figure T16.1 *Distributions of echo power and integrated power. There are 100,000 samples as for figure 3.6. (a) Noise distribution. (b) Distribution assuming that a 3-dB target has been added to the noise. (c) This time the target is 0 dB. (d) The noise distribution after video integration of 20 samples. Mean power is set at 1. (e) Noise + 3-dB target after video integration. The mean power is 3. (f) Noise + 0-dB target after video integration. The mean power is 2.*

two distributions of figure 2.5, separated by a carefully chosen threshold. In figure T16.1 the threshold must be placed at such a power level as to exclude as much of the noise distribution as possible, and to include as much of the "noise-plus-target" distributions as possible. Clearly, this placement can be made much better for the integrated distributions than for the original, un-integrated echo distributions. For example, placing the threshold at about 1.5 power units, for the integrated case, would eliminate most of the noise (low false alarm probability) but retain most of the noise-plus-target samples (high target detection probability). Obviously, the probability of detecting the 3-dB target will be higher than the probability of detecting the 0-dB target.

The threshold power is chosen ahead of time, based on detailed calculations of target and noise/clutter statistics. A train of echo pulses is then integrated, and a target is declared to be present if the integrated sample power

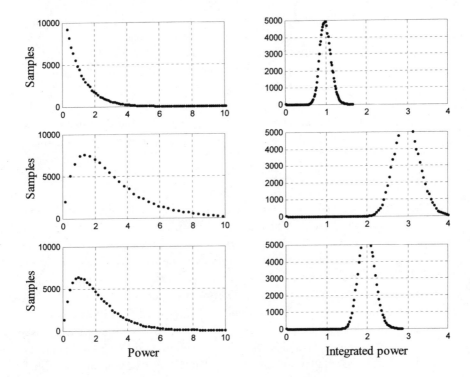

Figure T16.2 *Same as figure T16.1, except now 50 samples have been video integrated. Note that the distributions are narrower—hence easier to threshold.*

exceeds the threshold. The distributions in figure T16.1 tell us about the probabilities for detection and false alarm for our integration scheme.

Now consider figure T16.2. In this figure I have repeated the simulation of figure T16.1 except that this time I have integrated 50 pulses instead of 20. Compare the two figures and you see that the mean power levels have not changed: the noise + 3-dB target still sits at three, and the noise + 0-dB target still has mean power of two. This video integration behavior contrasts with that of coherent integration. In that case we saw that increasing the integration length (the number of echo signals added together) caused an increase in the SNR. Here it makes no difference. (Many nonspecialists, and even quite a few radar and sonar specialists who should know better, think that the purpose of integrating echo pulses is to increase SNR, and that only by doing so will target detection be improved. Not true, as we see here. Improved target detection does not require increased SNR.) Look, though, at the narrow integrated distributions compared with those of figure T16.1. You can see that setting a threshold is easier, and will result in fewer false alarms and more target detections. This is what I mean by improved detectability. If we increase the number of echo pulses in our video integration then the distributions become still narrower. Fewer pulses integrated lead to broader distributions and worse detection. (The three graphs on the left of figs. T16.1 and T16.2 assume no integration—or if you prefer an integration of one echo. Try setting a threshold to separate noise and target distributions here.)

Video integration works because of the nature of noise and target statistics. The distributions become narrower when we add together different samples (echo pulses), and so the "noise" and "noise plus target" echoes become more distinct.

T17. See Further with CFAR

Our airborne radar has scanned the ground below, looking for rats. I don't suppose that rat-detecting radar really exists—but I can pretend. These rats are infiltrating our farms and granaries, and are spreading havoc, despondency, and disease in roughly equal portions. In desperation, DARE has been called in—that's us—to search the farm and relocate any rats found near the Exclusion Zone. (DARE stands for Destroy All Rodent Enemies, "relocate" means "kill," and the Exclusion Zone is Farmer Fred's land.) Any rats that we detect from the air will be targeted by our ARMs (Anti-Rat Missiles). Trouble is, rats have a small RCS (rat cross-section) and Fred's farm contains a lot of different clutter sources. Before launching ARMs at anything that crosses our radar threshold (Fred might not like it if we were to zap his wife, who is feeding some chickens, and he would certainly not like it if any of his cows became collateral damage) we apply a CFAR.

Our radar transmitter scans Fred's farm. Each section of farmland, as it is illuminated by the radar main beam, returns echoes to our receiver. The clutter contained in each scanned section varies in power, because the land over which the beam travels contains different types of clutter—fields, trees, a pond—with widely varying cross sections. Our CFAR calculates the mean clutter power in each section, and sets the detection threshold at an appropriate level—say 15 dB—above the local mean power level. This avoids the problem discussed in the main text, arising if we apply a global threshold. Furthermore, a sophisticated CFAR can customize the threshold calculations for the local clutter. So, over the pond we find that clutter fluctuations about the mean value are small, so we can lower the threshold to 12 dB above mean. The wooded area contains clutter that fluctuates wildly, so to maintain a constant false alarm rate, our calculations show that we must raise the local threshold to 19 dB above the local mean.

I illustrate the CFAR scanning process in figure T17.1. One important feature you will see is that the scanned area (this is the area illuminated by our radar beam, at any given instant) has been tampered with by our CFAR processor. It has had a hole cut in it. So, when we calculate the local mean clutter

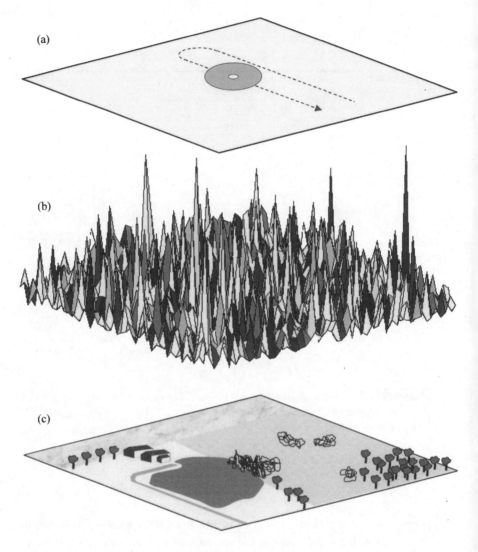

(a)

(b)

(c)

Figure T17.1 *(c) Fred's farm. (b) How our radar sees Fred's farm—most of the spikes represent strong local clutter sources. (a) The radar beam scans over the farm, and the constant false alarm rate (CFAR) processor calculates local threshold levels based on local clutter measurements.*

level, we use data from all of the scanned area except for this hole. The hole represents a single resolution cell, or perhaps a few such cells in a high-resolution radar system. We are testing this cell to see whether it contains a target, and estimate the mean clutter level in the cell from the *nearby* clutter. There is an assumption here: the clutter power in our "test" cell is the

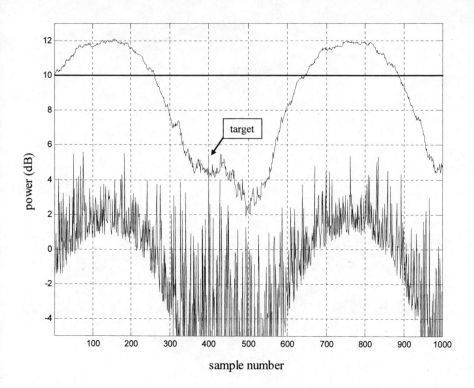

Figure T17.2 *Clutter (fluctuating lower lines) here varies with sample number. A weak 0-dB target has been included in sample no. 400. If, to ensure a low false alarm rate, we set a high global threshold level—here, 10 dB above the global mean clutter power (horizontal gray line)—then we miss the target. If we set a* local *threshold level, at 10 dB above the* local *mean clutter (black line), then we detect the target and reject the clutter.*

same as the mean clutter power in nearby cells. This assumption is statistically reasonable if the landscape does not vary very rapidly in local clutter power. Why do we bother with a test cell? Because of target *self-masking,* which I will now describe.

Suppose our test cell contains a target. The test cell thus stands out from its immediate neighbors that don't contain a target, and so we may hope to detect the target with an appropriately set local threshold. We must exclude the test cell from our estimate of local clutter power, however, because otherwise *the target itself would contribute to the estimated local mean clutter value.* We don't want this, because the target is a target and not clutter. We should not assume, when we test it, that a cell contains only clutter—there would be no point in testing. So we do not include the test cell in the measurement.

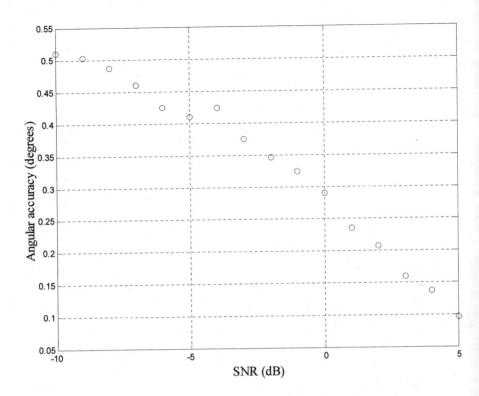

Figure T18.1 *The beam-matching sector scan algorithm improves the accuracy with which we estimate target direction. Here the radar beamwidth is $3°$ and I assume that there are five hits in the beam; you can see that accuracy is better than the beamwidth, and improves as target signal-to-noise power ratio increases. This simulation generates random samples of noise, and averages the result of 1,000 angle estimates for each point, using different noise each time. The graph is not a smooth curve because of statistical fluctuations, but the trend is clear.*

A further simulation serves to show how CFARs work. This time I will restrict the simulation to one dimension. You can see in figure T17.2 that a global threshold setting misses the target, which happens to be sitting in an area of weak clutter, but surrounded by areas of strong clutter. A local CFAR-calculated threshold undulates with the clutter, and so is able to pick out the target (target exceeds threshold) while rejecting the clutter.

Rats detected, ARM launched—squeak. Fred is happy. All because of CFAR.

T18. Sector Scan

In figure T18.1 I show the results of a computer simulation of the beam-matching sector scan algorithm, assuming that there are $n = 5$ hits in the

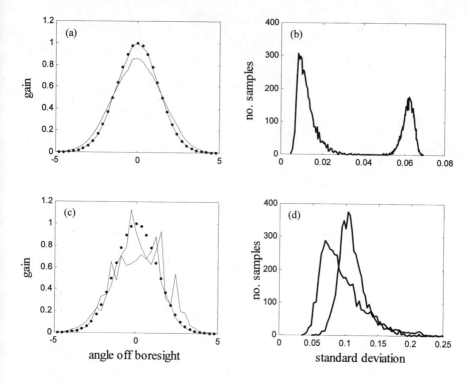

Figure T18.2 *The beam-matching sector scan algorithm applied to resolving two targets. Here the beamwidth is $\Delta\theta = 3°$, and there are 10 hits in the beam. (a) True beamshape is shown by the dotted line. Received signal powers are shown as solid lines, for one target present and—slightly broader—for two equal targets separated by 1.5°. The SNR is 15 dB. The extra width, due to the target separation, is discernible. (b) Standard deviation between the echo signal and the true beamshape, for one target and two targets. This is based on running a simulation 5,000 times, with different noise samples each on run. (c) Same as (a) but with SNR = 5 dB, for which the influence of noise is more obvious. (d) Same as (b) but for SNR = 5 dB. The one-target and two-target cases are less distinct.*

beam, and that the radar beamwidth is $\Delta\theta = 3°$. Even at low SNR there is a significant improvement in the accuracy of estimating target direction, but the algorithm really shines at high SNR. For these simulations I have assumed that the target is one that does not fluctuate in RCS, and so all the variability is here due to the effects of noise. In a real system the accuracy would flatten out at about 0.1°, no matter how high the SNR gets, because of other sources of statistical variability (for example, the influence of clutter, and of fluctuating targets). The beam-matching algorithm works a little better if we can increase the number of hits in the beam.

Angular resolution $\delta\theta$ can also be improved by this algorithm. For example, with the same 3° beamwidth we can distinguish two targets that are separated by only 1° or less, depending on SNR. Beam-matching works here because two targets that are close together in angle will appear as if they are one target detected by a slightly broader beam. The targets are not resolved by the beam, but they do cause it to appear wider than it really is. Because we know the true beamwidth, we can detect even quite small increases due to the presence of multiple targets. This is shown in figure T18.2, where the disprepancy in beamwidth appears as a standard deviation between real and echo beamwidths—another example of statistics entering the signal processing of remote sensors. In figure T18.2 I simulated 5,000 detections of one and of two targets and plotted the distribution of standard deviations. For large SNRs (and sector scan resolution improvement requires larger SNRs than does accuracy improvement) the standard deviations are distinct, whereas for lower SNRs the two distributions blur together. Such simulations show that two targets of equal RCS can be reliably separated if they differ in angle by only a quarter of a beamwidth, for a SNR of 15 dB. If the SNR drops to 5 dB, the targets can be distinguished when they are separated by about two-thirds of a beamwidth. So, resolution is improved, though not so dramatically as accuracy is, by sector scan processing. The improvement arises because we are making use of more information than is contained in the echo signal: we are applying our knowledge of the real transmitter beamshape.

T19. Correlation Processing

In figure T19.1 we have a sequence of echo pulses passing a copy of the transmitted pulse. (Think of a sprinter passing the finishing line, filmed with strobe lighting.) For clarity I have assumed that the echo pulses have not been mutilated by the environment—they return exactly as they were transmitted. When we correlate identical signals, the technical term is *autocorrelation*, and that is what we will do here. In practice, the echoes would be attenuated and cut up in various ways, so that we are comparing similar but not identical waveforms—this is so-called *cross correlation*. In this book, however, I will not draw this distinction, and both will be labeled "correlation."

Let us say that the echo pulse flies past our stored copy at time $t = 0$, exactly. At this time the two signals are perfectly aligned. We multiply them together, component by component (there are five components to the signal envelopes of figure T19.1), and add the components together. This process yields a maximum output value because the two signals are perfectly aligned. At other times the signals are not perfectly aligned and the multiply-and-add process yields a smaller value. The sequence of values, as a function of time, is our correlation function. It peaks when the signals align at $t = 0$, and is

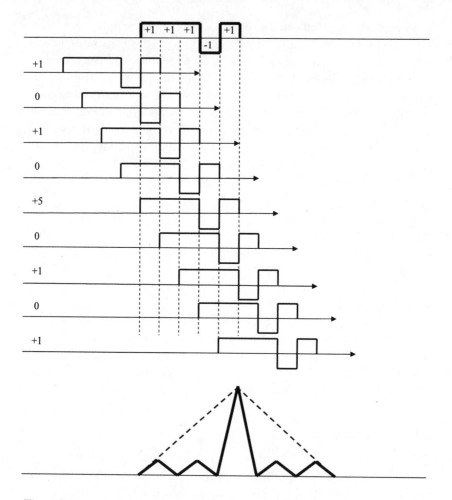

Figure T19.1 *A copy of the transmitted pulse envelope (bold, at top) is compared with an echo pulse at different times. As the echo passes by, each component is multiplied by the corresponding component of the copy, and all these products are added—the sum is shown on the left for different stages as the echo passes by. This correlation function is plotted at the bottom. Note how it peaks sharply when copy and echo are perfectly aligned. If we use an ordinary pulse envelope (all components +1) then we still get a peak at the time of perfect alignment (dashed line) but it is not so pronounced.*

smaller at earlier and later times. The correlation function is zero when the two signals do not overlap at all.

The length of the pulse envelope in figure T19.1 is 5 units (nanoseconds, feet, whatever) and so the natural accuracy of range estimation for this pulse is 5 units. With correlation processing, you can see that this accuracy is increased (the error in range estimation is reduced) to one unit. This result applies for the

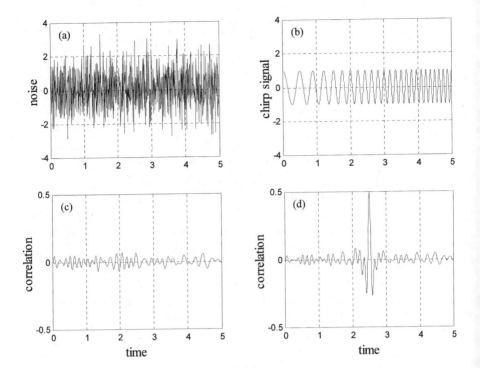

Figure T20.1 *(a) Noise samples, with a mean amplitude of 1. (b) Linear FM chirp pulse signal, with amplitude 1 and bandwidth B. (c) Correlation of the chirp with noise. (d) Correlation of the chirp with chirp-plus-noise, which represents a target echo. The width of the central peak is 1/B. Clearly, it is possible to pick out a chirp echo amid noise with high range resolution (high resolution due to the narrow spike).*

phase coding (sequence of +1 and −1) of figure T19.1 and for no phase coding at all (all components are +1). The phase coding makes it easier for us to see when the signals align, because it yields a correlation function with a pronounced peak when alignment is perfect. This pronounced peak is easy to detect and tells us the precise time at which our echo returns from the target.

Phase codes of this type (there are many variants) are known as *Barker codes*, and they have the property that, for a code of length n, the peak output is n, whereas all the other outputs are 0 or 1. The longest known Barker code has $n = 13$. However, range accuracies can be improved by factors much greater than 13 if we apply a more subtle and continuous phase variation along the length of each pulse. Cue frequency modulation, back in the main text.

T20. Chirp Pulse Compression

Suppose that we transmit a pulse that has a chirp waveform, as shown in figure T20.1. We listen for echoes, and correlate the incoming signal with a copy of this chirp pulse. In the simulation of figure T20.1 I have assumed a SNR of 0 dB, that is, the returning echo pulse competes with receiver noise of the same power. If no chirp is present in the receiver signal—no target echo— then our processor is correlating the chirp copy with noise only, and the resulting correlation function is shown. If, on the other hand, an echo is present in the receiver signal, then our processor is correlating the chirp copy with chirp plus noise, and the corresponding correlation function is also shown in figure T20.1. Note the prominent narrow central spike. Comparing with the noise-only correlation function, you can see that if an echo is present in the receiver signal then it stands out like a sore thumb, even for a low SNR.

We find the width of the central spike to be $1/B$, where B is the chirp bandwidth. Comparing this with the width (duration) of the original pulse, τ, we obtain a pulse compression ratio of $B\tau$. This is the same value as that obtained by performing the chirp processing in the frequency domain (measuring the frequency difference f_d, as discussed in the main text). So, the effectiveness of a chirp processor does not depend on which way we do the processing. In practice, one domain or the other may be more convenient, depending on the remote sensor construction and purpose, and both methods can be found in radar systems in operation today.

Chapter 4. Tactics: Skunks and Old Crows

T21. Waves in a Changing Medium

Waves change direction when they hit a surface. The most familiar example of this is *reflection*, shown in figure T21.1: a wave bounces off an impenetrable smooth surface (say, a mirror) at the same angle as it hit the surface. If the surface can be penetrated, then the wave *refracts* through the surface, as shown. Refraction is less familiar to many people, though we have all seen it. Think of what happens when you look into a clear lake or pond—the light is "bent" at the surface. Lastly, waves *diffract* past sharp edges and through small holes (which can be considered to be sharp edges). These three properties of waves are intimately related and are well understood by physicists. Diffraction is the most complicated of these three properties; I will not explain it here since it is of limited application in the radar and sonar field. This is because diffraction is insignificant—the wave rays are bent very little—if the wavelength is short compared with the edge size. For example, you cast a sharp shadow because the wavelength of light is much smaller than you are. On the other hand, water waves passing round a harbor pier are observed to bend, as in figure T21.1, because the wavelength is comparable to the pier size. Usually, in considering the propagation (traveling) of radar waves through air, we can ignore diffraction, and need consider only refraction and reflection.

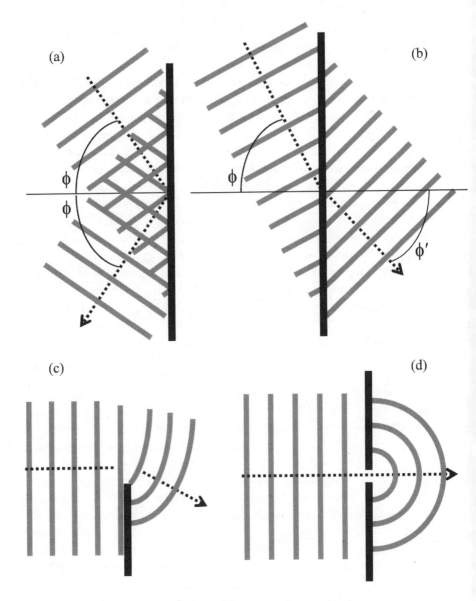

Figure T21.1 *(a) Reflection of waves from a surface (bold black line). The wavefronts (gray lines) and a ray (dashed line, with arrow) are shown. The ray is perpendicular to the wavefronts. Note that reflection angle is the same as incident angle. (b) Refraction of waves through a surface. The surface separates two regions of different wave speed. Here the waves move more slowly on the right side. (c) and (d) Two views of diffraction—the bending of waves around an edge.*

Reflection is so familiar that I need not waste your time with a physics text-book exposition, but refraction requires some kind of explanation. In figure T21.1 you see wavefronts passing through a penetrable surface—think of light moving through air and then through glass. At the glass surface, the light bends. We can explain this by imagining the wavefronts as lines of marching soldiers. The soldiers march easily on pavement, but move with more difficulty—more slowly—over heavy, softer ground. Say they march at a steady 4 mph on pavement and at a steady 3 mph over soft ground. In this analogy, pavement represents air, which light passes through easily, and soft ground represents glass, which, being denser, light passes through with more difficulty—more slowly. The point to note is this: the soldiers can maintain their lines when crossing the boundary from pavement to soft ground (from 4 mph to 3 mph) only if they shorten their step size, while maintaining the same rhythm. Otherwise the lines will fragment. The figure clearly shows that, given the change in step size, the soldiers' lines can be maintained across the boundary only if they change direction slightly at the boundary.

Light moves more slowly in glass than it does in air. The light wavefronts are maintained only when the wavelength becomes shorter in glass than it is in air (the light frequency is unaltered). Only in this way can the light penetrate through the boundary: it must change speed, and so it alters wavelength and direction.

We are not interested in light passing through glass, but in microwaves passing through the earth's atmosphere and sound waves passing through water. The atmosphere becomes less and less dense with increasing altitude, so that microwaves (and visible light and all EM radiation) pass through it more easily at high altitudes than at lower altitudes. I can model the atmosphere as horizontal layers of air, stacked one on top of the other, as shown in figure T21.2. The densest layer is near the surface, the next layer is a little less dense, and so on up. The boundary between any two layers is like the air–glass boundary in figure T21.1, and our microwaves change direction at the layer boundary in the same way. For the stack of layers, we obtain a microwave ray path as shown in figure T21.2—compare with figure T21.1 and you will see that the same thing is going on in both cases. If we make the horizontal layers very thin, then the ray path becomes a smooth curve— which is what we see. Light rays bend downward as they travel down through the atmosphere. They bend toward regions where the speed is slower. In practice the curvature of these rays depends on the wave frequency, but is always very small—the light travels almost in a straight line. However, over long distances the refraction effect can be significant and is responsible for a number of optical illusions—mirages. It also is a limiting factor in the resolution that can be attained by radar imaging systems, as we will see in chapter 6.

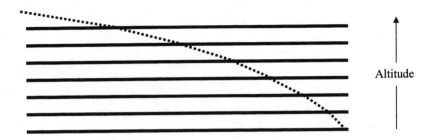

Figure T21.2 *Atmospheric refraction: a "layered" atmosphere, with each layer less dense than the one below it. A ray (dots) is refracted through each surface separating the layers. For the real atmosphere, this results in a curved path.*

Seawater becomes more and more dense with increasing depth. (Sound wave speed *increases* in a denser medium, whereas EM wave speed *decreases*—this is a significant difference between these two types of wave.) So sound waves bend as they travel deeper into the ocean. The refraction phenomenon is much more pronounced in this case—the bending is greater—because the variation in sound speed is greater. This is because sound speed changes with water salinity and temperature as well as water depth. Both temperature and salinity vary markedly, and so sound wave bending is more acute. Their bending goes the same way as for EM waves—toward regions of slower speed.

The formula that sonar engineers use to determine the speed of sound in water is:

$$c = 1449 + 4.6T - 0.055T^2 + 0.0003T^3 + (1.39 - 0.012T)(s - 35) + 0.017d$$

where c is the speed of sound expressed in m s^{-1}, T is temperature (°C), s is salinity (parts per thousand—the norm is 35 ppt), and d is water depth (m). From this formula we can show that sound speed increases at the rate of about 3 m s^{-1} for every 1°C increase in water temperature. Speed increases by 1.2 m s^{-1} for every increase in salinity of one part per thousand. Speed increases by 0.017 m s^{-1} for every meter increase in depth. The temperature and salinity gradients in ocean water are most pronounced near the surface, and so we expect that sound waves will bend more near the surface than they do at greater depths.

Acoustic refraction effects are so strong in seawater that they influence the tactics of submarines. Microwave refraction in the atmosphere is in general less important for radar operation, but reflection effects certainly influence tactics. These tactical consequences of anomalous propagation are discussed in the main text.

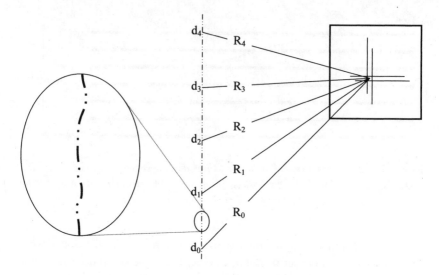

Figure T22.1 *"Range walk": the range changes as the radar antenna moves. We must keep track of these changes to build up a SAR image.* (Inset) *The airplane track is not truly straight: every little deviation must be monitored, to the nearest centimeter.*

Chapter 5. Mapping: Hearing the Picture

T22. Focusing SAR Images

Figure T22.1 is a magnified version of figure 5.2. The distances $R_{0,1,2}$ . . . from one particular pixel within the ground scene, to each of the synthetic array elements, are shown. Note that these distances change as the radar platform—the airplane—moves along the synthetic aperture flight path. This is a problem, becoming more acute as resolution improves, because the scene pixel under consideration will move through several radar range cells during the flight. So which range do we assign to the pixel? This problem is known as *range walk*, and is easy—though computationally intensive—to fix. For each and every pixel in the scene we say that the "true" range is from the pixel to the array element at the center of the synthetic aperture (say at d_2 in fig. T22.1), and adjust the echoes at other array elements so that they correspond to this range. This means that we take the echoes at $d_{0,1,3,4}$ and change their phases to the phases they would have had at the true range. It is easy to calculate what these changes are, and to implement them, but the difficulty is simply the number of computations necessary. If the scene is a square kilometer and the pixels are a square meter, then the scene consists of one million pixels. If the synthetic aperture is 1 km long and the synthetic elements are 20 cm apart, then there are 5,000 elements, and 4,999 need their phases adjusted for each pixel. Furthermore, the elements transmit and receive more than one pulse when imaging the scene, and these range-walk adjustments need to be made for each pulse. You get the idea. Happily, in these days of massively parallel computer processing, such computational tasks are well within current capabilities. However, it is not difficult to see why radar people claim that their signal processors have the highest data rate or throughput of any computers in the world.

Range walk is the coarsest focusing adjustment. The next level is the so-called *quadratic phase error* (QPE). This error is also a consequence of the SAR imaging geometry. Each of the pulses returned from each of the scene pixels will be Doppler shifted, because the pixel and radar platform are moving toward or away from each other. The Doppler shift is different for each pixel and this is crucial to forming the image—we do not want to get rid of it. However, the Doppler shift for any one pixel changes throughout the flight in a way that is not uniform. The SAR processing assumes that it *is* uniform—as if the platform was flying in a circle around the scene—and so the residual QPE must be phase compensated. (Roughly speaking, QPE is the radar equivalent of chromatic aberration in an optical lens.) Again, the calculations are easy and the phase correction required can be determined straightforwardly from the SAR geometry for a chosen scene. Again, the potential problem is the number of calculations and phase adjustments that need to be performed. And again, parallel processing comes to our rescue. QPE identified and corrected.

In figure T22.1 you can see that I have also magnified the airplane platform flight path. The flight path is not a straight line, but varies slightly, in a random and unpredictable way. This is because real planes are buffeted by turbulence and variable cross winds and so forth, so that they do not maintain a precisely straight flight path. The variations are small; perhaps the flight deviates from a straight line by no more than a few centimeters either way, but that is enough to destroy phase coherence for an X-band radar system, with a wavelength of 3 cm. (We want short wavelengths because they improve resolution, but this is one of the prices we pay.) So, the deviations from a straight-line flight trajectory need to be measured accurately and then compensated. The measurement is usually achieved by *accelerometers*, devices that are placed at various points on the airframe, and that measure acceleration at these points. From the reports of these accelerometers (which are becoming very accurate over a wide swath of acceleration rates) a flight control computer can piece together the airframe accelerations and twists—surge, heave, sway, pitch, roll, and yaw—and determine how much the radar transmitter and receiver have deviated from a straight line during the synthetic aperture flight. The appropriate adjustments are then made for the echoes of each element of the synthetic array. This is an impressive achievement, considering that the flight deviations must be corrected to within a fraction of a centimeter all along the line.

I hope it is clear that the computational problems of focusing a SAR image are large, and that they multiply exponentially as the required resolution increases. Now imagine what happens if something within the scene *moves* during the imaging period . . .

Focusing—phase correcting the echo pulses—works perfectly, in principle, only at the scene center. It works well for pixels that are close to the scene center and degrades further away. In practice this means that focusing works only over a certain area. Sometimes, in particular for SAR systems with moderate resolutions, the focused area is larger than the scene being imaged. For high-resolution images it is usually the case that the focused area is smaller, and this means that several different focusing corrections need to be made before the entire scene appears crystal clear—more processing.

The type of focusing that I have discussed so far (QPE) is a consequence of geometry, and so we need to know the geometry to focus the image. Other sources of defocusing are not analyzed so readily, and yet in some cases it is still possible to focus a SAR image. Nowadays there are several *autofocus* algorithms. These algorithms are becoming increasingly sophisticated, and are finding applications outside the field of SAR. Much of the image processing performed in your digital camera is autofocus. I will not try to explain the details, but simply note that the main feature of all autofocus techniques is that they work on the raw image. They do not need to know how the image was formed—the synthetic aperture geometry means nothing to them. They work entirely with the data of the blurred, defocused image and require no other input. Huh? Well, certain broad assumptions are made, such as that the defocusing progresses smoothly across the image. Then the algorithm looks for trends across the image that fit in with these broad assumptions. A model of the blurring is constructed, and model parameters are changed to minimize the blurring. The best parameter choice is converted into phase corrections, and applied to the image. Result: all defocusing that is compatible with the algorithm assumptions is minimized. Defocusing is often primarily due to one simple reason, such as QPE, and so an autofocus algorithm can identify and deal with it very well.

I am aware that my autofocus "explanation" is rather vague—blurred, even—but a more detailed explanation would not fit in with the rest of this book, in that I can find no simple geometrical explanation that can get you to the core of it without math. Apologies, but such is life.

Chapter 6. Special Applications and Advanced Techniques

T23. *Aliasing in Its Various Manifestations*

Aliasing is a characteristic—usually a problem but sometimes an asset—of digital signal processing. I will not discuss the wider consequences of this

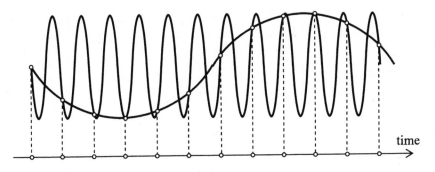

time

Figure T23.1 A *signal sampled digitally*, *that is*, *at discrete points*, *may not estimate frequency correctly*, *as you can see — more than one frequency is represented by the sample values*. *A theorem in DSP tells us that*, *to capture accurately all the frequency components of interest*, *we should sample at a rate which is at least twice the maximum signal frequency*.

phenomenon, but have already mentioned several aspects of it that are relevant to our subject. So, a brief aside to explain aliasing is long overdue. Say "aliasing" to a DSP (digital signal processing) hardware engineer and his or her first thought will likely be of the frequency ambiguity shown in figure T23.1. Typically a time domain signal is being sampled digitally, and aliasing raises the problem of unambiguously estimating the signal frequency, because digital sampling produces an ambiguity, as you can see in the figure. However, this aspect of aliasing is really an implementation problem, and is not the most important for us. Throughout this book I have been concerned with the algorithms of remote sensing, rather than with their hardware implementations, and so it is the consequences of aliasing for remote sensing techniques that I will concentrate on here.

I first mentioned the problem of range ambiguity way back in chapter 1 (see note T2, should you feel the need to refresh your memory). The dilemma is shown again in figure T23.2, along with a solution. You can see that range ambiguity really is an aliasing problem, since we have a digital sample (echo pulses) of a target, and consequently an ambiguous interpretation of the signal (ambiguous range). Range ambiguity does not arise in all radar applications — it depends on the chosen PRF and the maximum radar range. When it does occur, we can get around the problem by emitting pulses at two different PRFs. Say one burst contains five pulses and the next burst (of the same duration) contains seven pulses. From figure T23.2 it is clear that the target echoes will line up only at the true range. This system is not perfect. I will leave it to you to show that there is ambiguity every 35 pulses, but this may correspond to ranges that exceed the maximum radar detection range,

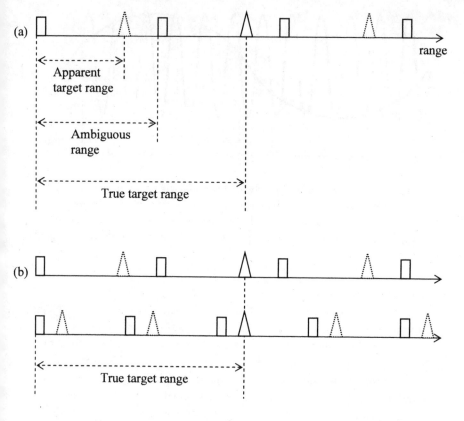

Figure T23.2   *Range ambiguity arises from pulse repetition, as we have seen. (a) How can we tell which is the true range? (b) We can determine the true range by transmitting pulses at different PRFs (pulse repetition frequencies).*

and so in practice the ambiguity will not arise. If it persists, then a third PRF can be added. (The mathematicians among you may realize that these PRFs should be "relatively prime." If not, then ambiguity occurs more frequently. For example, if instead of 5 and 7, I had chosen 6 and 8, then ambiguity would occur every 24 pulses, instead of 48. If prime numbers are not your thing then, as so often with remote sensing, the picture—figure T23.2—tells the story.)

A closely related consequence of aliasing was mentioned in footnote 9 of chapter 3, where I briefly noted the problem of "clutter foldover." A distant source of clutter (the moon, perhaps—see chapter 7) may be detected if it is particularly strong, and this clutter will "fold over" into closer range bins, as shown in figure T23.3. There are two bad consequences. First, this ghost clutter may mask the target, even though clutter and target are really at very dif-

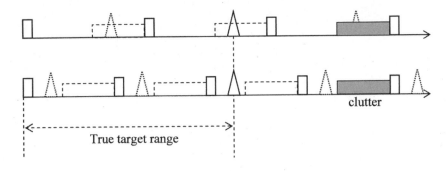

Figure T23.3 *Clutter foldover occurs when a strong distant clutter source (gray) shows up, because of aliasing, in closer range bins (dashed rectangles). These may mask the target, even though the clutter and target really are at different places.*

ferent ranges. Second, different sources of clutter can fold over and pile up on top of each other, producing what looks like very strong clutter at a short range. (A well-designed radar system will take this possibility into account.)

Aliasing is responsible for Doppler ambiguity as well as range ambiguity. We can see that this may arise from figure T23.1, if we recall that Doppler speed is measured as a frequency shift. A rather different consequence of aliasing is discernible in figure T4.1. Way back then I was considering the beamshape that results from linear arrays of transmitter elements. Well, a linear array of discrete elements is a spatial version of digital sampling and leads to aliasing problems. In the case of antennas we find that sidelobes pop up in the forward direction if the spatial separation of the elements exceeds half the transmitter wavelength, as in figure T4.1(b)—but not in figure T4.3, where the element spacing was less.

This last technical note is inserted almost apologetically, since it concerns a very technical aspect of remote sensing. Aliasing is ubiquitous, however, and tricky enough to catch even the experts, as you will see in chapter 7. We have seen that aliasing pops up everywhere but can be squeezed out into the margins by proper design practices. In the same spirit, aliasing pops up in this book, but has been squeezed out of the main text and into footnotes and technical notes. Live with it.

Glossary

Accelerometer   Device that measures acceleration.

Accuracy   The ability of a *radar* or *sonar* to identify a *target* position. Thus, a radar accuracy of 3 m means that the radar will estimate a target range to within 3 m of its true range.

Active   A *remote sensor* that transmits signals to the outside world and processes the echoes.

Aha!   I am, at last, undeceived.

Algorithm   A program, a formalized procedure or set of instructions.

Aliasing   Ambiguity that arises from digital sampling. The subject of note T23.

AM   Amplitude modulation (of a *waveform*).

Amplitude   The height of a wave (the pressure of a sound wave, or the voltage of an oscillating electrical signal). Peak amplitude is related to wave energy.

Amplitude monopulse   An *algorithm* for refining direction estimates. The subject of note T6.

Anomalous propagation   Bending of rays through the atmosphere, due to spatially varying atmospheric density.

Antenna   An array or dish that concentrates transmitted or received power in a chosen direction.

ASDIC   Allied Submarine Detection Investigation Committee. An early (British) name for sonar.

ASW   Anti-submarine warfare.

ATC   Air traffic control.

Attenuation   Loss of transmitted power due to absorption or scattering by the atmosphere (or by water, in the case of underwater sonar). Discussed in note T10.

Autocorrelation   *Correlation* of a *signal* with itself.

Autofocus   An *algorithm* that can focus a *remote sensor* image without reference to the data-gathering technique.

Azimuth   Horizontal direction angle. Thus, north and east differ by 90° in azimuth. Also known as *bearing*.

Bandwidth (BW)   The range of frequencies needed to describe a *waveform*, or the range of frequencies that a signal processor can process. The subject of note T7.

Battlespace   The volume of space above a battlefield.

Beamforming   The process of constructing a transmitter or receiver beam from constituent elements that are usually *isotropic*. The subject of note T4.

Beamrider   A dumb missile that simply follows a beam to a *target*.

Beamshape   The form of an *antenna* beam; the dependence of transmitter power on angle off *boresight*.

Beamwidth   The angular width (in *azimuth* and *elevation*) of a *radar* or *sonar* beam.

Bearing   *Azimuth* angle.

Bin   A term used by systems engineers to describe a cell, a small element of a larger range of values. Thus, the frequency *spectrum* is divided into many frequency bins, and the range of a *radar* may be divided into a large number of range bins. Bin size almost always corresponds to *resolution* cell size.

Bistatic   General term for *remote sensors* that have separate *transmitters* and *receivers*.

Blanking   Plugging ears. Intermittently shutting off the *receiver* so that it is not damaged by the nearby high-powered *transmitter*.

Blinking   A method of deceiving *home-on-jam* missiles.

Boresight   The direction that a *radar* or *sonar antenna* is pointing. Thus a *target* is said to be "on boresight" when the antenna is pointed right at it.

Bright band   A weather *radar* feature due to ice melting in the atmosphere.

Burn through   The act of penetrating through a *jammer signal* to detect an underlying *target*.

Cavitation   Violent implosion of underwater bubbles—can lead to submarine propeller damage.

Cell   A division of *radar resolution* space. Also known as "bin" or "box."

CF   Constant frequency. A simple *remote sensor waveform*.

CFAR   Constant false alarm rate. A CFAR processor seeks to maintain the same *threshold* level, so that the *false alarm* rate does not fluctuate.

CH(L)   Chain Home (Low). The first radar system to be deployed successfully. "Low" refers to a fix that eliminated a blind spot at low *elevation* angles.

Chaff   Artificial clutter (thin metal strips) disbursed in the air to blind enemy *radar*.

Channel   Independent processing chain. A tracking *radar* with 100 channels can process 100 *targets* simultaneously.

Chaotic   A physical system with behavior that depends sensitively on starting conditions. Also, the way in which very large technical projects are managed.

Chirp   A chirp *waveform* is one that changes frequency rapidly with time.

Clutter   Unwanted power reflected back into a *radar receiver* from the environment.

Clutter canceling   A technique for reducing the *clutter* content of a *remote sensor* echo, when the *target* is moving.

Coherent   A *waveform* is coherent if the *phase* difference between adjacent pulses is known, otherwise it is noncoherent. A processor that measures only of the return waveform power, and not the waveform phase, is said to be noncoherent.

Conformal array   A *phased array* that is shaped to conform with an airframe.

Controlled air space   The volume of air space monitored by the world's ATC controllers. All other air space is "uncontrolled."

Cooperative target   A *target* that wants to be detected.

Corner reflector   A reflector that sends radiation back to its source, whatever the source direction.

Correlation   A mathematical measure of similarity, in our case, of two *waveforms*.

Countermeasures   (CM) Tactics adopted by a *target* to evade detection or destruction. *Electronic countermeasures* (ECM) are a part of *electronic warfare* (EW).

Counter-countermeasures   (CCM) Tactics adopted by a *radar* or sonar to avoid being spoofed by *target countermeasures*.

Cross correlation   *Correlation* of one *signal* with another signal.

Cross-eye   A *jammer* that transmits a *signal* that deceives a *radar receiver* about its direction of origin.

Crossed dipoles   A simple form of *transmitter* or *receiver antenna* that permits direction estimation.

CW   Continuous wave. A *remote sensor* that transmits without interruption.

DBS   Doppler beam sharpening. A *SAR* variant.

Decibel (dB)   A logarithmic ratio of two values (e.g., the Richter scales for comparing earthquake magnitudes). Convenient when measuring vastly different magnitudes. Discussed in note T8.

Decorrelation   The changing of a *remote sensor target* or *clutter signal* with time.

Deep sound channel   A layer beneath the ocean surface that transmits acoustic radiation a long distance. A *duct*.

Diffraction   The bending of waves around sharp edges. See note T21.

Dipping sonar   A *sonobuoy* is lowered into the ocean from a helicopter, and later retrieved.

Doppler   Pertaining to *target* speed relative to the *radar/sonar*. A Doppler processor measures such speed; a Doppler shift is the change in frequency of an electronic *waveform* reflected from a moving *target*. Discussed in note T15.

DSP   Digital signal processing

Duct   A layer of water or atmosphere that traps sound waves within it.

Dynamic range   Extent of a *receiver's* sensitivity.

Echo   See *return*.

Echolocation   Detecting a *target* position using sound waves. An echolocator is an organic *sonar* (a bat or a whale). Also known as biosonar.

ECG   Electrocardiogram. A *remote sensor* specializing in imaging the heart.

ECM   Electronic *countermeasures*.

ECCM   Electronic *counter-countermeasures*.

Elevation   Direction angle measured up from the horizon, or from horizontal. Straight up is 90° elevation.

EM   Electromagnetic.

Emergent behavior Here, the capability of a network that arises from the network itself, rather than from the inherent capability of individual network constituents.

Ensonify Illuminate with sound.

Envelope The magnitude of a *waveform*, the maximum amplitude.

Ergonomics The study of efficient working arrangements.

EW Electronic warfare.

False alarm *Noise* or *clutter* power is mistaken for a *target*, for example, by exceeding a *threshold*.

Feeding buzz The *FM waveform* adopted by bats during the terminal phase of an attack sequence.

Flow noise *Noise* resulting from an object moving through a fluid.

FM Frequency modulation.

Footprint The shape of an airborne *transmitter* beam on the ground.

Fourier transform Here, a mathematical transformation that takes a *waveform* from the time domain to the frequency domain. An "inverse Fourier transform" converts a frequency domain waveform back into a time domain waveform.

Frequency Referring to a wave, this is the number of *amplitude* peaks that pass a given point per second.

Gain The magnification of *transmitter* power (or of *receiver* sensitivity) in a particular direction, due to an *antenna*. Gain is expressed relative to *isotropic*. See note T9.

Hits in the beam The number of pulses from a rotating *antenna* that reflect from a *target*.

Home-on-jam Technique of *beamrider* missiles that follow a *jammer signal* back to its source.

Hydrophone Underwater microphone.

IFF Identify friend or foe.

Incoherence A property of many textbooks and legal documents.

Integration Adding together echo signals to improve *target* detectability.

Interference *Clutter* plus *noise*—the unwanted component of a receiver waveform.

Ionospheric Reflection Bouncing radio waves off the ionosphere to see *OTH*.

IR Infra-red.

ISAR Inverse synthetic aperture radar.

Isotropic The same in all directions. So an isotropic *antenna* has no favored direction.

Jammer Transmits noise into a receiver to prevent *remote sensing*.

K, Mr. An *entirely fictional* character who messes up complex remote sensor projects through self-promotion.

Klystron A source of *coherent* microwave radiation.

Lloyd's mirror A *sonar* synonym for *multipath*.

Loss factor A catch-all fudge factor that summarizes atmospheric attenuation of propagating *EM* waves.

Magnetron A device that efficiently generates centimetric *microwave* energy.

Mainlobe The central portion of an *antenna* beam, defining the *boresight* direction.

MAV Micro-unmanned air vehicle.

MB800   Facetious term used in this book to describe *echolocating* bats.

Mean value   Average.

Microwaves   That portion of the *EM spectrum* that is usually transmitted by *radar* systems.

Monostatic   *Remote sensor* that has colocated *transmitter* and *receiver*.

MTI   Moving target indicator—a *radar* mode of operation.

Multipath   More than one pathway exists for a *signal* to be transmitted, reflected, and returned to the *receiver*.

Multistatic   *Bistatic remote sensor* with more than one *transmitter* or *receiver*.

Neyman-Pearson criterion   Statistical criterion for judging the validity of a hypothesis.

Noise   Unwanted electronic *waveform* that corrupts the desired *signal*. Noise originates inside the *radar* or *sonar*, and so does not depend on where the antenna is pointing or on the outside environment. White noise is characterized by having the same amplitude in every frequency bin.

Noncooperative target   A *target* that does not want to be detected.

Noncooperative transmitter   A *transmitter* co-opted for use by a *remote sensor* receiver, but which was designed for a different purpose.

OTH   Over the horizon.

Passive   A *remote sensor* that receives but does not transmit.

PD   *Pulse Doppler*. A *remote sensor* that incorporates coherent signals.

PDF   Probability density function. Measures the probability that a sampled random variable will assume a particular value.

Phase   Describes, along with *amplitude*, the state of a wave.

Phase coding   *Phase* modulation of a transmitted *signal*, to aid signal detectability through *correlation* processing.

Phased array   An *antenna* consisting of array elements that transmit with a controlled *phase* relationship between elements.

PRF   Pulse repetition frequency.

PRI   Pulse repetition interval.

Probability   Frequency of occurrence.

Pulse   A brief burst of transmitter power.

Pulse compression factor   The improvement in range resolution due to *chirp* or *phase-coded* signal processing. See T20.

Pulse train   A rapidly repeated series of transmitted pulses.

QPE   Quadratic phase error. A source of *SAR* image defocusing. Discussed in note T22.

Radar   RAdio Detection And Ranging. An active *remote sensor* that applies EM radiation, usually *microwaves*.

Range-gate pull-off   A *repeater jammer* seduction technique.

Range walk   A source of *SAR* image defocusing. Discussed in note T22.

Ray   A line denoting the direction of travel of a wave.

RCS   Radar cross section. This is the effective area of the *target*, as observed by the *radar*. I use the same term for *sonar* targets. Discussed in note T11.

RDF   Radio direction finder. An early (British) term for *radar*.

Receiver The *antenna* that gathers up a *radar* or *sonar* return and feeds the return *waveform* into the *signal* processor. Ears are acoustical receivers.

Reflection Bouncing of a wave off a surface. See note T21.

Reflectivity field A two-dimensional display of atmospheric echo power.

Refraction Change of direction of a wave passing through a surface boundary separating regions of different density. See note T21.

Refractive index Measure of the optical density of materials through which *EM* waves travel.

Remote sensor Uses waves to gain information about, or to form an image of, the environment.

Repeater A sly *jammer* that deceives a probing *radar*.

Resolution The minimum difference that can be detected by a *radar* or *sonar* sensor. Two *targets* in the same resolution cell will be detected as a single target. If they are in the same range bin, but in adjacent azimuth bins, then the targets are "resolved in azimuth." The subject of note T5.

Return The signal + clutter that is reflected back to the *radar* or *sonar* receiver from the environment. Received power. Synonymous with *echo*.

Reverberation *Sonar* equivalent of *clutter*.

Rope Long *chaff*, designed to blind long-wavelength radars.

Route control The *ATC* controllers who monitor the *controlled air space* between airports.

ROV Remotely operated vehicle.

RWR *Radar* warning receiver.

Sampling Measuring a *signal* at a particular time or sequence of times.

SAR Synthetic aperture radar. The method used by humans to form images from *radar* data. The radar platform is moving, and forms a large (synthetic) *antenna*.

Scan rate The rotational speed of an *antenna*.

Scene The area imaged by a *SAR* or *sidescan sonar*.

Sector scan *Antenna* repeatedly scanning a particular region of space. Also an *algorithm* for improving angular *accuracy* or *resolution*.

Self-masking Describes a bad *algorithm* for *target* detection, whereby target *signal* power contributes to target obscuration.

Sensor fusion The intelligent application of different *remote sensor* types (*radar*, sonar, optical, *IR*) to learn more about a distant *scene* or *target*.

Sidelobes The (unwanted) portions of an *antenna* beam, distinct from the *mainlobe*.

Sidescan *Sonar* equivalent of *SAR*.

Signal A *radar* or *sonar* return, due to reflections from the *target* and the environment. Sometimes the term is restricted to the target echo.

SNR Signal-to-noise ratio. The ratio of *target* power to *noise* power in a *radar* or *sonar signal* processor.

SOJ Stand-off *jammer*. A transmitter that sends *noise* into an enemy *radar receiver* to blind it and so protect an ally vulnerable to radar.

Sonar SOund Navigation And Ranging. An active *remote sensor* that applies acoustical radiation (sound), usually at high frequencies.

Sonobuoy A disposable *sonar* sensor that is dropped by aircraft into the ocean.

Sonographer An echocardiogram operator.

SOSUS SOund SUrveillance System.

Spectrogram A two-dimensional plot of *waveform amplitude* and frequency.

Spectrum The frequency components of a *waveform*. More precisely, the distribution of waveform power in different frequency *bins*.

Spot jammer A *noise jammer* that concentrates its output into a narrow bandwidth.

Spotlight The high-resolution version of *SAR*.

SSJ Self-screening *jammer*. A *target* that transmits *noise* into an enemy *radar receiver* to blind the radar and so protect itself.

Standard deviation Measure of variation of samples from a distribution.

STAP Space-time adaptive processing. New-fangled *clutter* rejection technique.

Stripmap The low-resolution version of *SAR*. Covers a lot of ground quickly.

Superposition An important property of waves: when two of them overlap they simply add up in amplitude. Sometimes called *wave interference.*

Target What the *radar* or *sonar* is looking for.

Terminal control Airport *ATC* controllers, who monitor the volume of space near airports.

Threshold A power setting to decide whether or not a *signal* contains a *target*.

Towfish A towed *sonar array*, used for forming *sidescan* images. These images may or may not involve a synthetic aperture.

Transducer Underwater acoustical *transmitter.*

Transmitter The power source plus *antenna* that emits *microwave* energy into the air (*radar*) or acoustical power into the air or water (*sonar*).

Transmitter blanking Switching off a *transmitter* so that a *receiver* can hear the echoes.

TWT Traveling wave tube—a source of coherent *microwave* radiation.

UAV Unmanned (sometimes the more politically correct "uninhabited") air vehicle.

Velocity field A two-dimensional display of atmospheric velocity.

Video integration Summing echo powers to improve *target* detectability. Also known as noncoherent integration.

Waveform The shape of an electronic pulse.

Wavefront A line of constant *phase*, for example a line along the crest of a wave. A wavefront is perpendicular to the direction of motion of the wave, and so to the *ray*.

Wavelength The distance between contiguous peaks of a wave.

WMD Here your author, who seeks to explain his subject Without Mathematical Details.

Index

Bat (*continued*)
processing, 142–53; spectrogram, 144, 148; waveforms, 140
Battle of Britain, 8, 26–37; aircraft, 27–29, 29n, 31–34; losses, 27, 29, 30; participants, 27n, 33; significance, 35–36
Battlespace, 183n
Beam: mainlobe, 206–7, 210; nulls, 208; sidelobes, 206–7; steering, 174–76, 176n
Beamforming, 178, 207–11
Beamshape, 46
Beamwidth, 84, 124, 126, 136, 155, 206–7; broadening, 176n
Bell Labs, 14, 217
Bin. *See under* Resolution
Biosonar. *See* Echolocation
Bistatic radar, 168–74; advantages, 172–73; disadvantages, 173–74
Blinking, 100
Blip, 1
Blitz, 33–37, 102–3
Boresight. *See under* Antenna
Box. *See under* Resolution
Bright band, 160
British: CH radar development, 13, 14, 15–22, 38, 41; ECM, 105; H2S navigation system, 41, 43, 105
Brookner, Eli, 195
Burglar alarm, 187–88
Butterfly effect, 158
Buzz, 1; feeding (*see under* Bat)

Cathode ray tube, 167
Cavitation, 57
Cetacean. *See* Whales
CFAR (constant false alarm rate), 79–81, 242–45
Chaff, 94–96; flares, 97; RCS, 95
Chain Home (CH): CHL, 22, 23; radar (*see under* Radar); stations, 17–21, 23, 30; towers, 17n, 18
Chaos, 158
Chirp ranging, 86–89, 147–49; FM (*see under* Waveform)
Clutter, 52–55, 68, 122, 228–32; canceling, 71–72; correlation, 75–76; filter, 68, 71; foldover, 81n, 258–59; ground, 70, 229; sea, 60n; simulation, 69, 71, 80; spatial variability, 69–71;

79–80; suppression, 70–73, 180; surface, 54, 228–30; volume, 53–54
Cold War, 112, 116–18, 230
Computational load, 123, 254–56
Conferences, 194n
Conical scan, 85, 101n
Consortia, 193–94
Copulating shrimps, 6, 57, 86, 164
Corner reflector, 229–30
Correlation, 85–86; processing, 247–49. *See also under* Clutter
CW waveform. *See under* Waveform

DARPA (Defense Advisory Research Project Agency), 186
Daventry experiment, 14
Decibel (dB), 47, 217–20
Deep sound channel, 111
Depth charges, 113
Diffraction, 250–51
Dolphins. *See under* Whales
Domain: frequency, 73, 181; time, 73
Doppler: beam sharpening (DBS), 129; effect, 67, 70, 149, 237–39; processing, 65–73, 96, 123, 143–52, 159, 239; and relativity, 237; shift, 71, 89, 132, 144, 171, 180, 238
Dowding, Hugh, 24, 29–30, 36, 39
DuBridge, Lee, 9, 39
Dwell time, 83
Dynamic range, 219

Earth curvature, 22, 107–8
Echo, 73, 86, 87, 231; fluctuations, 60, 215, 234–35
Echocardiogram, 167–68, image, 169
Echocardiology, 166–68
Echolocation, 1, 4, 58, 136; non-mammalian, 136. *See also* Bats; Whales
Ejection fraction, 168
Electrocardiogram (ECG), 168
Electromagnetic (EM): HF, 15, 17, 22n, 108; microwave radiation, 4, 38, 92–94, 108, 223; spectrum, 4, 91, 99, 221–22; visible, 108; wave attenuation in atmosphere, 222–24; —in seawater, 55–56; wave scattering, 22n, 223–24; wave speed, 10, 10n, 12n, 238

Moth model, 145; spectra, 146, 151
MTI (moving target indicator), 72–73
Multipath, 107, 108–9

Navigation: bird, 67; systems, 41–44, 102, 105
Nessie, 166
Neural networks, 187, 189n
Newhouse, R.C., 14
Neyman-Pearson criterion, 62
Noise, 51, 53, 57, 171; external, 57
NRL (Naval Research Lab), 9n, 11, 14, 38

OTH (over-the-horizon), 82, 108

Page, Robert, 14
Park, Keith, 26–27, 30, 36n
Passive: jamming, 94–96; sensing, 5, 94
PDF. See under Probability
Pearl Harbor, 15, 39–41
Phase. See under Waves
Phased array. See under Antenna
Ping, 1
Ponte, Maurice, 9
Power. See under Radar; Sonar
PPI (plan position indicator), 43, 159, 160
PRI. See under Pulse
Probability, 232–33; density function, 234–35
Processing load. See Computational load
Pulse, 81, 125, 204–5; burst, 234, 236; coherence (see under Waveform); compression factor, 88, 249–50; repetition frequency (PRF), 83n; repetition interval (PRI), 18, 204, 236

Quadratic phase error, 255–56

Radar, 1; acronym, 4, 4n; altimeter, 14; applications, 44; bistatic, 168–74; capabilities, 66, 90; Chain Home (CH), 2, 14, 15–22, 41, 214, 216; CH system, 22–26; clutter limited, 55, 148; coherent, 66, 72, 88, 235–37; conferences, 194n; consortia, 193–94; death ray, 13n; development, 22, 24, 24n; display, 121–22, 159; image, 130; invention 9–11; noise-limited, 55; non-coherent, 72; power, 17; pre-war, 11–16; pulse-Doppler, 66n, 72, 239; pulses,

19–21; ship navigation, 10; signal power, 50–51; synthetic aperture (SAR), 122–31, 153; tracking, 75, 83, 85, 221; weather, 68, 157–61
Radar warning receiver (RWR), 96, 98, 106
Radome, 178, 178n
RAF. See Royal Air Force
RAM (radar-absorbing material), 226
Randall & Boot, 38, 42, 43
Range ambiguity. See Ambiguous range
Range equation, 48–51; loss factor, 224
Range-gate pull-off, 100
Range walk, 254–55
RCS (radar cross section), 48, 53, 84, 224–28; fluctuations, 60, 85, 225; reduction, 226–28
RDF (radio direction finder), 4n
Receiver: bandwidth, 216; beamshape, 84; multistatic, 171–72; update rate, 83
Reflection, 250–51. See also under Corner reflector
Reflectivity field, 160
Refraction, 127, 128, 128n, 250–53
Remote sensing, 2, 2n, 6, 57, 90, 123, 127; geometrical explanation of, 3; IR and optical, 123n
Resolution, 127; vs. accuracy, 211; angular, 15, 123n, 124, 211–14, 246–47; cell, 64, 81–82, 124–25; cross-range, 123–26; range, 87, 124, 147, 151
Return. See Echo
Reverberation. See Clutter
Rope. See under Chaff
ROV (remotely operated vehicle), 182–87
Royal Air Force (RAF), 16, 24, 31, 33, 36, 105

SAR (synthetic aperture radar): DBS, 129, 131; focusing, 254–56; inverse (see under ISAR); scene, 125; shadow processing, 130, 190; spotlight, 125, 129; stripmap, 125, 129. See also Radar
Scan rate, 82–83
Scottish, Scots, 2n, 9, 13, 18; Isles, 193
SEAD (suppression of enemy defenses), 106
Sector scan, 82–85, 143, 245–47
Sensor: fusion, 142n, 187–90; multiplication, 188

Shadow zone, 110
Signal: correlation, 85–86; filtering, 68;
Silent Sentry, 172
Simulations: bandwidth, 217; bat, 143n,
146; CFAR, 244; chirp pulse
compression, 249; clutter, 69, 71, 80;
echo power distribution, 240–41;
integration, 74; noise, 52; sector scan,
245–46; statistics, 233; video integration,
78; wave, 202
SIR (signal to interference ratio), 65, 78
Skolnik, Merrill, 9
SNR (signal to noise ratio), 51–52, 74–75,
77, 241
Sonar, 1, 11, 11n, 55–58, 76, 110–11;
acronym, 4, 4n; applications, 134n;
array, 133; coherent, 88; fish-finding,
164–66; image, 135, 166, 169; mirrors,
11n; operator efficiency, 77; power, 111;
sidescan, 122, 133–36; vs. radar, 57–58,
93–94
Sonobuoys, 113, 113n, 118n
Sonographer, 168
SOSUS, 118
Sound speed: in air, 200; in seawater, 56,
253
Spectrogram, 144
Spectrum, acoustic, 146–48. See also under
Electromagnetic
Spotlight SAR, 125–26, 129
Standard deviation, 233–34
STAP (space-time adaptive processing),
181–82
Statistics, 59–62, 232–35; clutter, 59;
target, 59
Stealth aircraft, 172n, 226–28; image, 227
Stripmap SAR, 125, 129
Submarine, 56, 57, 110, 112–16, 119;
ballistic missiles, 116–17; German (see
under U-boats); hunter-killer, 117n,
118n; nuclear, 116; Soviet, 116, 116n,
119; tactics, 119–20; United States, 119;
wake, 117
Surface duct, 111
Synthetic aperture, 125–28

Target, 4n; classification, 66, 145–47; 155,
188; cooperative, 66n; decorrelation, 75;
fish, 165; location, 117; noncooperative,

66n; phase, 73; range, 87; resolution,
212; self-masking, 244; signal, 62, 73–78;
speed, 68, 70
Taylor & Young, 11–12
Tesla, Nicholas, 11
Test cell, 243–44
Threshold, 61, 79, 187, 240, 242; local, 81,
244
Tizard, Henry, 13–14, 32; mission, 9,
37–38, 92
Tomography, 152
Towfish, 133, 134–35
Transducer, 56, 167
Transmitter, 125, 136; beamshape, 83–84;
blanking, 88n, 142; CH, 19–21, 170;
footprint, 182; Klystron, 48, 236;
magnetron, 38, 39n, 48, 236; power, 16,
17, 48–50, 153; TWT, 48, 236
Transponder, 163

UAV (unmanned air vehicle), 183–85;
autonomous, 185; cost, 185n; Firescout,
184–85; Predator, 183–85
U-boats, 41–42, 113; effectiveness, 113–16;
losses, 114
Ultra, 36, 37, 37n
United States: Cold War ASW, 117–18;
ECM development, 104–6; Loran
navigation system, 44, 108; radar
development, 11–14, 15, 22,
38–39, 43
Update rate, 83

Velocity field, 160

Walker spy ring, 58n
Warfare, electronic, 73
Watson-Watt, Robert, 13–15, 16, 170
Waveform: bat, 140; CF, 144; chirp FM,
86–88, 144, 147–50, 165, 180, 249–50;
coherent, 235–36; CW, 11–13, 171;
noncoherent, 235; whale, 153
Wavefront, 101, 109n, 203
Wave interference. See under Waves
Wavelength, 4, 44, 199–200
Waves, 4, 199–204; amplitude, 199–200;
bending (see under Anomalous
propagation); coherent, 66; crest, 201,
207; envelope, 204–5; frequency, 199;